T0314277

INTRODUCTION TO TOPOLOGY

PRINCETON MATHEMATICAL SERIES

Editors: Marston Morse, H. P. Robertson, A. W. Tucker

1. The Classical Groups, Their Invariants and Representations. By Hermann Weyl.
2. Topological Groups. By L. Pontrjagin. Translated by Emma Lehmer.
3. An Introduction to Differential Geometry with Use of the Tensor Calculus. By Luther Pfahler Eisenhart.
4. Dimension Theory. By Witold Hurewicz and Henry Wallman.
5. The Analytical Foundations of Celestial Mechanics. By Aurel Wintner.
6. The Laplace Transform. By David Vernon Widder.
7. Integration. By Edward James McShane.
8. Theory of Lie Groups: I. By Claude Chevalley.
9. Mathematical Methods of Statistics. By Harald Cramér.
10. Several Complex Variables. By S. Bochner and W. T. Martin.
11. Introduction to Topology. By Solomon Lefschetz.

INTRODUCTION TO TOPOLOGY

By SOLOMON LEFSCHETZ

PRINCETON, NEW JERSEY · 1949

PRINCETON UNIVERSITY PRESS

PRINTED IN THE UNITED STATES OF AMERICA

LONDON: GEOFFREY CUMBERLEGE
OXFORD UNIVERSITY PRESS

Preface

The present work originated from a short course delivered in 1944 before the Institute of Mathematics of the National University of Mexico. Lectures on a number of the topics dealt with in the book were also given from time to time at Princeton University. Generally the auditors had ample training in algebra and general topology but little in algebraic topology. While this topic was to some extent emphasized in the lectures, it was found advisable to provide considerable background material. This has also been the author's purpose throughout the present volume.

The main topic presented here is the topology of polyhedra, and its treatment begins properly with the third chapter. The indispensable preliminary survey is meant to give a bird's eye view of topology without much regard to the specific subjects presented later. The first chapter, on foundations, contains practically no proofs. Chapter II stands apart from the main text. It is selfcontained and deals only with dimensions at most 2 and chains mod 2. A surprising number of the basic concepts shows up already at that early stage so that the chapter is an excellent introduction to the rest of the book.

Here and there throughout the volume, but above all in the first chapter, proofs of certain propositions are omitted, a fact indicated by an asterisk before the statements.

A sparse bibliography will be found at the end of this book. The references are given by the author's name followed by a specific indication in square brackets (the caps refer to books and the lower case letters to papers).

From Chapter II on, and except for a few problems, the only spaces considered are polyhedra. Armed with this knowledge the reader will be well prepared for the more advanced and more complete treatises.

The author wishes to express here his grateful appreciation for many valuable suggestions to Dr. H. F. Tuan, of the National Tsing Hua University, to Mr. Jaime Lifshitz, of the National University of Mexico, who read the manuscript, and to Dr. E. E. Floyd and Mr. H. W. Kuhn who read the galleys.

S. Lefschetz

Contents

INTRODUCTION TO TOPOLOGY

Introduction: A Survey of Some Topological Concepts

The concepts which it is proposed to examine in this survey are above all those related to algebraic topology. Little regard will be paid to historical matters as the existing and very accessible bibliographies make it easy to trace most questions under discussion to their sources.

1. Theory of Sets. Topological Spaces.

Topology begins where sets are implemented with some cohesive properties enabling one to define continuity. The sets are then called topological spaces. Thus topology is a branch of general set theory, the creation of Georg Cantor (around 1880). That his ideas have had a profound influence in all mathematics is well known. In topology this influence has been decisive.

The cohesion giving rise to continuity may be obtained in various ways, notably by means of limits, closure relations, distances, or finally by the assignment of a collection of subsets: the open sets, or the closed sets, satisfying certain structural axioms. The choice of axioms is dictated by the behavior of the open or closed subsets of Euclidean spaces. For the choice made must be such that among the selected spaces are included the classical spaces. The topological spaces of various kinds began to come into their own through Hausdorff's classical treatise: *Grundzüge der Mengenlehre* (1912). The most important topological spaces are the metric spaces first introduced in their full generality by Fréchet (Paris thesis 1906), together with a number of other important notions, notably the notion of compactness. A discussion of these concepts will be found in Chapter I.

Once continuity is solidly established one may define continuous transformations or functions, henceforth called mappings. One may also define therefore bicontinuous transformations, also called topological transformations or homeomorphisms. Whatever such transformations maintain invariant constitutes the subject matter of topology proper.

In many questions, notably in algebra and in analysis, one has occasion to topologize sets of some sort and sometimes even in different ways for the same set. In particular a group $G = \{g\}$ may thus be topologized. Whenever this is done in such manner that the basic group operation gg'^{-1} is continuous in both g and g', G is said to be a topological group. Generally, but not always, some supplementary conditions are imposed. A noteworthy case is the additive group $\mathfrak{P}$ of the reals mod 1. This group is usually topologized by considering its elements as angles mod 2π and assigning the topology of the circumference to the system. Fundamental research has been made on these groups in recent years by L. S. Pontrjagin, whose theory of characters and group duality is classical. (See his book: *Topological Groups.*)

Much interest attaches to mappings as such. If A,B are fixed spaces one may consider classes of mappings $f_t : A \rightarrow B$, varying continuously (in a reasonable sense) with a real parameter t. Any two such mappings are said to be homotopic, and also to be in the same homotopy class. These homotopy classes may be conceived as topological characters of the pair of spaces (A,B), and thus justify (if this were necessary) the attention paid to them in topology.

Dimension is certainly one of the most important concepts of topology and its interest goes far beyond the bounds of our subject. The need for a clear-cut definition of the concept was first pointed out by Poincaré (1912). He proposed the following recursive definition: A space is n-dimensional whenever its walls are $(n - 1)$-dimensional; the null-set has the dimension (-1). Soon after L. E. J. Brouwer pointed out that Poincaré's definition was imperfect since it assigned the dimension 1 to a doubly sheeted cone. Brouwer also gave the first admissible modern definition. The one now generally adopted is due to Menger and Urysohn, the true (independent) creators of what is known today as dimension theory. Their definition goes somewhat as follows: A metric space A is of dimension n at a point x, written $\dim_x A = n$, if x may be isolated from the rest of A by an arbitrarily small set of dimension $< n$, where n is the least integer with this property; the dimension of A, written $\dim A$, is $\max \dim_x A$ for all points x of A, and is infinite when $\dim_x A$ is unbounded; the recursion is completed as before by assigning the dimension -1 to the null set. This definition assigns the "correct" dimensions to all the well-known sets.

A noteworthy property associated with the dimension of a space goes back to Lebesgue who based upon it a proof of the topological invariance of the dimension of Euclidean spaces. Stated for any metric space A it is as follows: Let A be decomposed into overlapping open sets and let $\mathfrak{U}$ be any such decomposition. The order of $\mathfrak{U}$ is the integer

n defined by the condition that there are $n + 1$ sets of the decomposition, but no more, which have a nonvoid intersection. Let $n(\epsilon)$ be the least n corresponding to all decompositions $\mathfrak{U}_\epsilon$ into sets of diameter $< \epsilon$. If A is metric and separable (i.e. has a countable dense subset) then dim A is max $n(\epsilon)$. This property is the one which ties up most closely with algebraic topology.

2. Questions Related to Curves.

The endeavor to grasp the true content of the term "curve" and to arrive at a satisfactory definition has had a very considerable influence upon the development of topology. It has also been the point of departure for many highly important and interesting investigations.

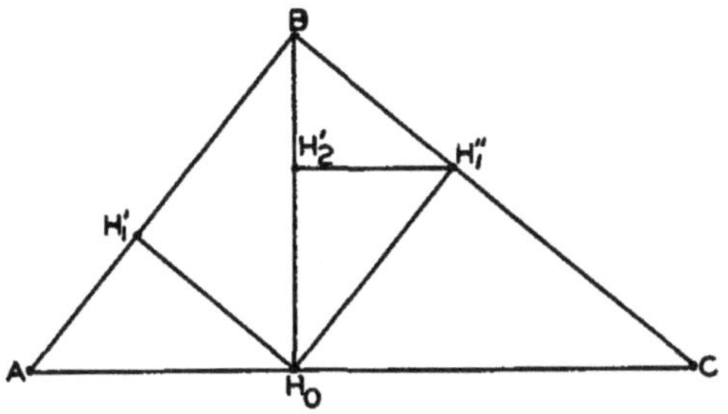

Figure 1

Until comparatively recently no one raised any objection to the analytic (nongeometric) definition of a curve, say in the plane, as a locus represented by a system

$$(1) \qquad x = f(t), \qquad y = g(t)$$

where f,g are continuous on the segment $0 \leq t \leq 1$. More briefly, then, one would have said: a curve is the continuous image of a segment. It was pointed out however by the mathematician G. Peano (1890) that a curve in the sense just described could fill up a square. It is a little simpler to show, by a construction due to Polya, that it may fill up a triangle. Consider a right triangle ABC with $AB \neq CB$. Given any $t_0 : 0 \leq t_0 \leq 1$, write it as a dyadic fraction $t_0 = 0 \cdot a_1a_2a_3 \cdots, a_i = 0$ or 1. Drop the perpendicular BH_0 from B to AC, then the perpendicular H_0H_1' to AB the smaller side of the right triangle ABC if $a_1 = 0$, the perpendicular H_0H_1'' to BC otherwise. Repeat the construction from H_1' or H_1'' as the case may be, with ABC replaced by AH_0B in the first case, by BH_0C in the second, etc. Call $H_1,H_2, \cdots$, the successive points reached (thus H_1 is H_1' or H_1''). The sequence $\{H_n\}$ thus arising from t_0 has for limit some point P_0 in the closed triangle ABC and we set $f(t_0) = P_0$. It is not difficult to show that $f(t)$ is continuous and that every point P of ABC is obtained as an $f(t)$. Hence f maps the segment $0 \leq t \leq 1$ onto the closed triangle ABC.

A curve in the general sense just discussed, is known as a continuous curve, also as a Peano continuum. What are then the characteristic properties of such a locus? Since the segment is a continuum, i.e. it is compact, metric and connected, and since these properties are preserved by a mapping, a continuous curve must certainly be a continuum. It was proved by Hahn and Mazurkiewicz (1914) that to completely characterize a continuous curve the only other necessary property is local

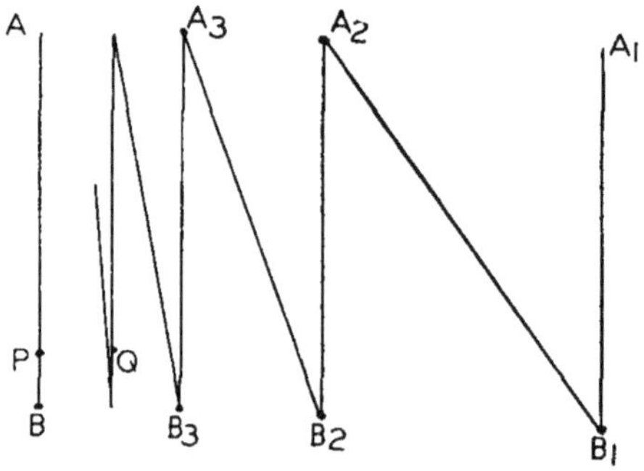

Figure 2

connectedness. This property may be described in the following manner: A space A is locally connected whenever given any neighborhood U of any point x (= open set containing x) there is another V contained in U such that any two points of V may be joined by an arc in U.

By virtue of the theorem of Hahn-Mazurkiewicz a cube, a sphere, a square, are all continuous curves. On the other hand, and this is a classical example, the planar continuum of Fig. 2 where $A_n \to A$, and $B_n \to B$, fails to satisfy the condition of local connectedness at any point such as P in Fig. 2. For there are points Q of the set arbitrarily near P which may not be joined to P by any arc in the set. Thus our set is not a continuous curve. For references on local connectedness and related questions notably the work of Borsuk, Kuratowski, Lefschetz, see the author's monograph, *Topics in Topology*, Annals of Mathematics Studies, No. 10.

It is plain that to define a curve as the continuous image of a segment is unsatisfactory, since it makes a curve of the square but rejects the set of Fig. 2 which one would naturally describe as a curve (of some sort).

At this point dimension theory came to the rescue with the definition: a curve is a metric space which is one-dimensional in all points. The square is known to be two-dimensional; the set of Fig. 2 is easily proved to have the dimension one in all points. Hence under the new definition the former is not a curve but the latter is.

To conclude with continuous curves we may observe that a considerable number of geometers have pursued investigations into their structure, among them many Polish topologists and R. L. Moore and his school.

Does a simple closed curve J (the homeomorph of a circumference) separate the plane in the way that it appears to do, i.e. into two regions

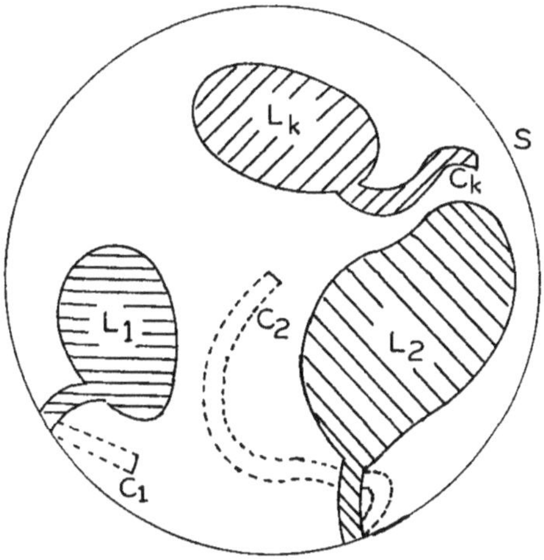

Figure 3

with J as their common boundary? The question was first raised by Camille Jordan (hence the name Jordan curve given to J), but it was *not* solved by him. Indeed this theorem has had a great record of incorrect proofs. A number of correct proofs are now in existence and one (correct we hope) will be found in Chapter II. The Jordan curve theorem has been noteworthy also as a starting point for fruitful research. L. E. J. Brouwer first gave a relation for the number of regions in which a plane is partitioned by a bounded closed subset. He also gave the extension of the Jordan curve theorem to the partition of an n-space by an $(n - 1)$-sphere. The full extension culminated in the duality theorem of J. W. Alexander (1922) which related the Betti numbers (definitions later) of a closed set A in a sphere S^n to those of the complement $S^n - A$. A definitive theorem on this question was given by Pontrjagin (1932).

One would be tempted to think that the Jordan curve property of being the common boundary of the component regions of its complement is a property possessed only by Jordan curves. Brouwer showed, however, that there is a large category of sets behaving in the same way. A very simple construction of such a set was given by the Japanese geometer Yoneyama. On a sphere S consider k lakes (regions) $L_1, \cdots, L_k$ filled with as many distinct fluids. Draw from each lake a canal and let L_i' be the union of L_i and its canal. If d is the diameter of S, let the canals be so drawn that every dry point of S (i.e. not in any L_i') is not farther than $d/2$ from each L_i'. The operation is repeated indefinitely with $d/4, d/8, \cdots$, so that at the n-th stage the dry points are each not farther than $d/2^n$ from each of the enlarged lakes. The set B of the dry points left is closed and is the common boundary of each of the k ultimate enlarged lakes.

3. Polyhedra.

Owing to their simplicity, their ready visualization and the ease with which they lend themselves to experimentation, polyhedra have naturally attracted the attention of topologists. From our point of view however "polyhedron" must be taken in a broad sense; curved faces are as acceptable as flat faces. For example, we would call a triangulated sphere a polyhedron. The class of objects thus singled out is therefore very rich. A deeper and very important additional reason for investigating polyhedra lies in a fundamental result due to P. S. Alexandroff (1928), according to which every compactum A may be indefinitely approximated by a polyhedron, and even by one of the same (but not of lower) dimension as A itself. This gives hope for carrying over, by suitable limiting processes, topological properties of polyhedra to compacta. This process has been in fact extensively utilized by Alexandroff and others. The best instance perhaps is the Čech homology theory for topological spaces.

Returning to polyhedra we may think of a number of topological properties, for instance the dimension and the number of connected parts. Others, much less obvious and essentially algebraic in kind, were discovered by Poincaré (1895–1901). They are tied up with his homology theory which is perhaps the most profound and far reaching creation in all topology. The guiding purpose is to obtain properties associated with the disappearance, or the reduction in dimension, of certain parts as a consequence of continuous deformations. Thus on a sphere all closed curves may be deformed into a point, but on a torus this is untrue for meridians and parallels. Similarly neither circumference nor the sphere are deformable on themselves into points. Homology will make it possible to catch, in a certain measure, these rather elusive properties.

Let us assume that the polyhedron Π is made up of a finite number of noninterpenetrating faces. This assumption does not constitute a topological restriction. We suppose the faces to be cells, the p-cells being written E_i^p. Their collection K is of a type known as a complex,

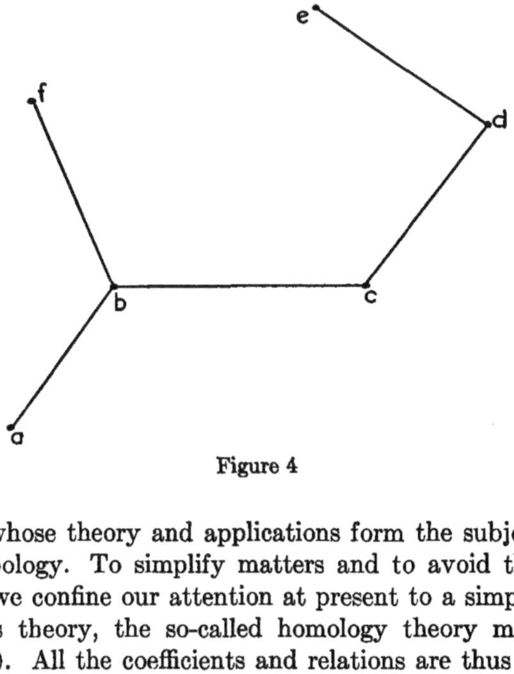

Figure 4

a structure whose theory and applications form the subject matter of algebraic topology. To simplify matters and to avoid the pitfalls of orientation, we confine our attention at present to a simplified version of Poincaré's theory, the so-called homology theory mod 2 due to Tietze (1908). All the coefficients and relations are thus to be understood at present to be residues mod 2.

The basic concept is the chain. A p-chain of K is simply a linear form $C^p = \Sigma g_i E_i^p$, little more than a collection of p-dimensional faces. (Later, when more general coefficients are to be considered the chain will be a "weighted" collection.)

Just as one may associate a boundary with a set so one may associate a boundary with the chains. For C^p it is a $(p - 1)$-chain $FC^p = \Sigma h_j E_j^{p-1}$, where $h_j = 1$ whenever E_j^{p-1} is a face of an odd number of E_i^p

occurring in C^p (i.e. with $g_i = 1$) and $h_j = 0$ otherwise. Thus in Fig. 4, drawn for $p = 1$, the vertices a,b,e,f are in FC^1 but the vertices c,d are not. In other words $FC^1 = a + b + e + f$. The chain C^p is known as a cycle if $FC^p = 0$ (all coefficients h_i are zero mod 2). Intuitively the chain may be thought of as "any polyhedron," the cycle as "a closed polyhedron." An FC^p is always a cycle. Such a cycle is called a bounding cycle.

If we think of F as an operator it is readily seen to be linear. Moreover the remark just made about bounding cycles means in substance that $FFC^p = 0$ whatever C^p. This is often written in operator form as

$$FF = 0. \tag{2}$$

This operator relation is the central property as regards our present argument.

From an intuitive standpoint FE^p is "deformable" over E^p into a point. Recalling our earlier intuitive remarks, on geometric grounds FE^p, and therefore owing to linearity, FC^p should be more or less neglected. We introduce then with Poincaré a homology $\gamma^p \sim 0$ to signify that γ^p bounds. (This is the origin of "homology theory.")

One may be tempted to think that all cycles bound. Consider however a torus triangulated in such manner that a meridian is a union of arcs and vertices of the triangulation. This meridian gives rise to a one-cycle $\gamma^1 \nsim 0$. Similarly for a suitable parallel. The sum of the triangles is a 2-cycle γ^2 and it is $\nsim 0$ since there are no 3-cells in the triangulation.

We must now launch more fully into algebraic considerations. Consider a linear combination $\Sigma g_i \gamma_i^p$, where the γ_i^p are cycles. The maximum number of p-cycles which have no bounding linear combination is called the p-th Betti number mod 2, written R_2^p. Thus for the torus $R_2^1 = 2$, $R_2^2 = 1$. If there are α^p cells E_i^p then $\chi(K) = \Sigma\,(-1)^p\alpha^p$ is known as the characteristic of K and we have the basic relation

$$\chi(K) = \Sigma\,(-1)^p\,\alpha^p = \Sigma\,(-1)^p\,R_2^p. \tag{3}$$

Thus for a convex polyhedron in ordinary space $R_2^0 = R_2^2 = 1$, $R_2^1 = 0$, hence $\chi(K) = \alpha^0 - \alpha^1 + \alpha^2 = 2$, the well known Euler relation (already known, however, earlier to Descartes) whose generalization (3) is due to Poincaré.

When the polyhedron Π is an n-dimensional compact manifold (smooth like a Euclidean n-space in all points) there takes place the Poincaré duality relation $R_2^p = R_2^{n-p}$, whose ultimate extension and clarification (for chains with arbitrary coefficients) is due to Pontrjagin (1932).

If one allows oriented elements and more complicated types of coefficients in the chains, other homology characters make their appearance. In particular the chains with integral coefficients give rise to new homology characters, the torsion coefficients. These denote the fact that there exist cycles γ^p which do not bound but of which a multiple $m\gamma^p$ may bound. A simple example is the polyhedron arising from a triangulated real projective plane $\mathfrak{P}$. The subdivision may be so chosen as to decompose a given line into a finite number of intervals. Their sum (duly oriented) is an integral one-cycle γ^1 which does not bound but whose double bounds.

A fundamental property of the homology characters is their topological invariance. Surmised by Poincaré, it was first proved by J. W. Alexander (1915) for the Betti numbers, then by Veblen (1922) for the other characters.

The topological invariance of the numbers R_2^p implies that they are independent of the cellular decomposition of the polyhedron. It follows also that the characteristic $\chi(K)$ is a topological invariant. It is in fact a very easily computed invariant of a polyhedron.

As first pointed out by Emmy Noether, the proper and only adequate formulation of the relations between chains, cycles, $\cdots$, requires group theory. The chains C^p form an additive group $\mathfrak{C}^p$. The cycles γ^p form a subgroup $\mathfrak{Z}^p$ of $\mathfrak{C}^p$ and the chains FC^{p+1} a subgroup $\mathfrak{F}^p$ of $\mathfrak{Z}^p$. The operator F induces a homomorphism of $\mathfrak{C}^{p+1}$ into $\mathfrak{C}^p$ whose kernel is $\mathfrak{Z}^{p+1}$ and the image $F\mathfrak{C}^{p+1} = \mathfrak{F}^p$. The factor group $\mathfrak{H}^p = \mathfrak{Z}^p/\mathfrak{F}^p$ (group of the p-cycles reduced modulo the bounding p-cycles) is the p-th homology group mod 2. Under the present assumptions $\mathfrak{H}^p$ is a vector space over the field of the residues mod 2, and its dimension is the number R_2^p.

The relationship just described between the groups $\mathfrak{C},\mathfrak{Z},\mathfrak{F},\mathfrak{H}$ is entirely typical and will recur constantly in the sequel. Thus, if one takes account of orientations in a certain standard way, one will have integral chains, or chains with other categories of coefficients (details will be found in Chapter III) but the $\mathfrak{C}, \cdots, \mathfrak{H}$ scheme will remain the same. A systematic study of such systems independently of any relation to topology has been made by Walther Mayer (1929 and on).

An important generalization of geometric nature (Lefschetz 1928) consists in neglecting throughout the cells in a certain subcomplex L of K. This gives rise to the homology theory mod L, or "relative" homology theory, the starting point for all the "relative" theories. Intuitively there is a close analogy here with the relative concepts of point sets. The relative and absolute theories are special cases of a theory of "abstract" complexes investigated at length by A. W. Tucker and the author. This more general theory may be made to include the recent "cotheories" (for cochains, cocycles, · · ·) to which all the extensions of Poincaré duality may now be related (Lefschetz, Alexander, Whitney, Čech). A few words regarding these concepts may not be amiss. Returning to our polyhedron Π, let us attach to each E_i^p a dual element E_p^i and reverse the incidences as follows: whenever E_j^{p-1} is a face of E_i^p we now say "E_{p-1}^j has the face E_p^i." The dimensional "going down one unit" is preserved by assigning to E_p^i the negative dimension $-p$. The boundary is computed by the same principle as before and the basic relation (2) continues to hold. The mutual boundary relations are

$$FE_i^{p+1} = \Sigma\eta_{ij}(p)E_j^p, \qquad FE_p^j = \Sigma\eta_{ij}(p)E_{p+1}^i. \tag{4}$$

The customary scheme used by most authors differs from the above in that they have only chains (no cochains, i.e. they continue to write E_i^p where we have E_p^i) and distinguish between the two operations in (4) by denoting them respectively by the symbols ∂, δ.

Using the second boundary relation in (4) as a basis, one defines new groups $\mathfrak{C}_p$, · · ·, $\mathfrak{H}_p$ exactly as before.

A mapping f of a polyhedron Π into a polyhedron Π_1 induces a mapping φ (homomorphism) of the homology groups of Π into those of Π_1 and a dual mapping φ^* of the cohomology groups of Π_1 into those of Π. Moreover, the two operations φ,φ^* are orthogonal in a certain sense and the situation bears a deep analogy to the well-known covariance-contravariance relation occurring so often whenever one deals with linear transformations. The full implications of the concepts here alluded to can only be developed within the framework of Pontrjagin's group duality theory, a theory avoided in the present work.

As a consequence of the φ,φ^* association cohomology has played a very important role in many recent developments (the work of Lefschetz on fixed points, of Whitney and Stiefel on sphere bundles and the imbedding of manifolds, etc.). There are also most suggestive contacts with integration to be considered later.

The homology characters have proved powerful enough to classify completely the compact manifolds of dimension two (see Chapter II) but they failed signally for the higher dimensions. This problem is one which greatly preoccupied Poincaré but is as yet unsolved. It is known for instance that an orientable closed surface is topologically a two sided disk with a certain number p of holes, the characteristic being then $2 - 2p$. No such models have ever been found even for three dimensional manifolds. Poincaré showed that two 3-manifolds may have the same homology characters and yet have distinct fundamental groups (see below) so that they are not homeomorphic. J. W. Alexander showed (1918) by an example that they could have the same homology characters and the same fundamental group, yet fail to be homeomorphic. And there the case lies.

Another noteworthy unsolved problem on manifolds is the following likewise proposed by Poincaré: Is every topological n-manifold (locally Euclidean space) a polyhedron? For $n = 2$ this has been established by Tibor Rado (1922). He was able to do so because of the accurate information at hand regarding the relation of a plane Jordan curve to its complement. This knowledge is largely absent however when it comes to higher dimensions. The best information available is due to R. L. Wilder.

Our ignorance regarding the classification of manifolds has not prevented the expansion of their theory otherwise. We may mention in particular: (a) an extensive theory of intersection of cycles (Lefschetz 1926) with many applications, notably, to algebraic geometry; (b) Whitney's modern theory of differentiable manifolds (1937) (the usual manifolds of differential geometry and mathematical physics).

The endeavor to classify manifolds led Poincaré to introduce another noteworthy concept, his group of paths, which lies at the roots of many of the difficulties of topology and as a consequence has been a powerful stimulus for new research.

Let us take an arc-wise connected compactum $\mathfrak{R}$ and let $\alpha, \beta, \cdots,$ denote paths in $\mathfrak{R}$, beginning and ending at a fixed point P of the space. Write α^{-1} for α reversed and $\alpha\beta$ for α followed by β. Write also $\alpha = 1$ if it is deformable (with end points always in P) into P itself. As a consequence the set $G = \{\alpha\}$ becomes a (generally) noncommutative group, abstractly independent of P. This is the fundamental group of $\mathfrak{R}$.

A noteworthy generalization of Poincaré's group is found in the homotopy groups of Hurewicz (1935). They arise by replacing the closed paths by images of a p-sphere S^p with a fixed point Q going into P. These groups, for $p > 1$, are all commutative and have stimulated much new research.

Significant related results are Hopf's theorems (1929) on the mappings of spheres on spheres. To mention only one, Hopf showed (1931) that there exists a mapping f of a 3-sphere S^3 onto a 2-sphere S^2 such that no continuous modification of f may free any point of S^2 from the image fS^3. For other dimensional combinations there are partial results due to Hopf, Pontrjagin and Freudenthal, but they are rather scattered and much remains to be done in this direction. The general problem is certainly difficult and calls perhaps for radically new methods.

4. Coincidences and Fixed Points.

Does a given transformation f of a set A into itself possess fixed elements? That is to say, is there in A an element x such that $fx = x$? One recognizes in this formulation the typical existence problem of analysis. Thus if Δ is any functional operator (differential, integral, both combined, etc.) and φ is a function, the solution of $\Delta\varphi = 0$ is equivalent to the search for a function φ invariant under the operation $1 + \Delta$.

The problem of coincidences is very similar: given two transformations f,g of a set A into a set B one asks if there exists an element x of A such that $fx = gx$. This problem has been extensively investigated for algebraic transformations of algebraic curves into one another. The fixed point problem corresponds to $B = A$ and g the identity.

For a segment $l : 0 \leq x \leq 1$ and its continuous transformations into another segment $m : 0 \leq y \leq 1$, there is a very simple and suggestive approach to our problem. Namely in the plane x,y consider the square $0 \leq x, y \leq 1$. Each transformation will be represented by an arc: λ for f and μ for g (Fig. 5). Their coincidences correspond to the intersections (x_i,y_i) of λ,μ. The transformation $x \to fx$ of l into itself may be represented in the square by λ, the identity mapping of l on itself by the diagonal ν, and the fixed points of f by the intersections of λ with ν. This model is the best possible guide when dealing with all questions of this nature.

Another interesting special problem deserves mention. Let $F(z)$ be a polynomial in the complex variable z. The relation $fz = z + F(z)$ defines a mapping of the complex plane into itself whose finite fixed points are the roots of $F(z)$. The author's fixed point theorem (see below) yields then a ready proof that such points do exist. To prove their existence is to prove the theorem of Gauss on the zeros of polynomials, i.e. the so-called fundamental theorem of algebra.

The problem of fixed points is then of evident interest. Looking at it purely topologically we first encounter (around 1910) Brouwer's fundamental theorem: Every mapping of a solid sphere (closed spherical region) into itself possesses a fixed point. A very intuitive proof may be given by means of vector fields. Let H^n be the solid n-sphere, f the mapping, x any point of H^n. If $y = fx$, the segment xy defines a vector field on the whole of H^n. The places where the vector is zero are the fixed points. The very simple theory of the index of vector fields (see below) shows that such points always exist and Brouwer's theorem follows. Brouwer also obtained partial results for spheres, projective spaces and rings. The author obtained (1926 and later) a much more general result to which a large number of the known fixed point theo-

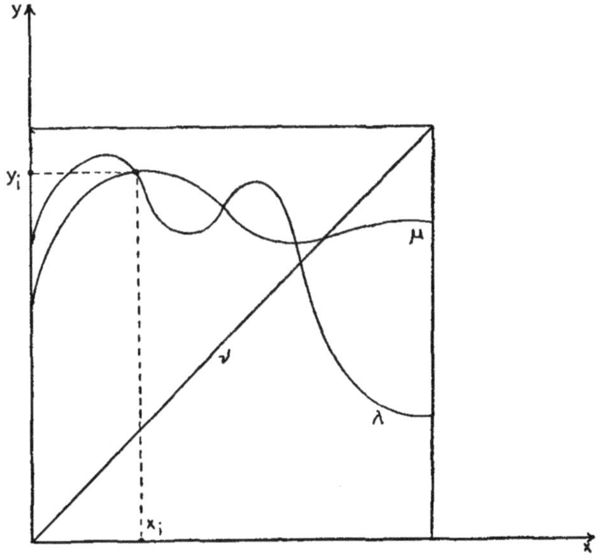

Figure 5

rems may be reduced. Let Π be a polyhedron giving rise to a complex K and f a mapping of Π into itself. For any dimension p let $\gamma_1^p, \cdots, \gamma_r^p$ be a maximum set of cycles of which no linear combination bounds. (The cycles may be rational or modulo any prime.) Then, assuming the extension of the homology concepts in a reasonable sense, f gives rise to a set of basic homologies

(5) $$\varphi\gamma_i^p \sim \Sigma\varphi_{ij}(p)\gamma_j^p \text{ in } K.$$

It is easily proved that the expression

(6) $$\Phi(f) = \Sigma(-1)^p\varphi_{ii}(p)$$

does not depend upon the selection of maximal sets $\{\gamma_i^p\}$. Moreover if the cycles are rational, $\Phi(f)$ is an integer. And now we may state: If $\Phi(f) \neq 0$, then f has a fixed point. Thus for mappings of the solid sphere into itself we always have $\Phi(f) = 1$, hence f always has a fixed point. This is Brouwer's theorem. Similarly for a sense-preserving [sense-reversing] mapping f of a sphere S^{2n} $[S^{2n+1}]$ into itself $\Phi(f)$ is always $\neq 0$. Hence such a mapping always has a fixed point (Brouwer).

There are fixed point theorems which escape the $\Phi(f)$ theorem. The most famous is the theorem of Poincaré-Birkhoff (surmised by Poincaré in 1912 and proved the same year by Birkhoff) regarding certain homeomorphisms of a plane annular ring. A plane ring bounded by two concentric circles Γ,Δ is subjected to an area preserving homeomorphism f into itself which moves all the points of Γ in one direction and all those of Δ in the other. Then the theorem asserts that there are at least two fixed points and this in spite of the fact that here $\Phi(f) = 0$.

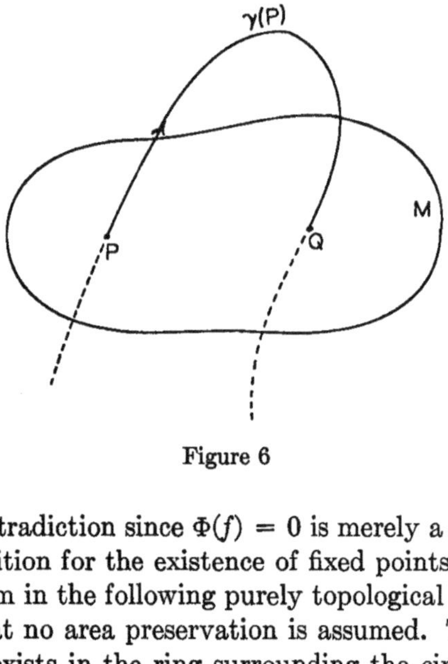

Figure 6

This is not a contradiction since $\Phi(f) = 0$ is merely a sufficient but not a necessary condition for the existence of fixed points. Birkhoff generalized the theorem in the following purely topological form: Let f be as before except that no area preservation is assumed. Then either some Jordan curve J exists in the ring surrounding the circle Γ which does not meet its image fJ, or else there are two fixed points.

Poincaré's interest in fixed points arose out of dynamics. Under certain conditions the trajectories of a system behave like a set of stream lines in the phase space or space of the coordinates and velocities. It may happen also that in that space there is a manifold of section M: a manifold which each trajectory keeps intersecting repeatedly. Let P be a point of M and $\gamma(P)$ the trajectory through P. As $\gamma(P)$ is followed beyond P in the direction of increasing time it will have a first intersection with M beyond P in Q. Thus Q is a function $f(P)$ of P which maps M into itself and its fixed points correspond to the periodic orbits of the system. These periodic orbits are of course of prime interest in many questions, notably in celestial mechanics, and they were very extensively investigated by Poincaré.

Considerable work on fixed points of certain periodic transformations has been done (since 1934) by P. A. Smith. The motivating question is to determine to what extent does a periodic homeomorphism of the n-sphere S^n, or of an Euclidean space $\mathfrak{E}^n$ resemble an orthogonal transformation. In particular is it equivalent to an orthogonal transformation? The full conjecture appears to be very difficult to verify. However, Smith has shown that regarding many homology properties and prime power periods, the conjecture is correct. Thus the set of fixed points of such transformations acting on S^n has more or less the homology properties of S^r, $r < n$, while acting on $\mathfrak{E}^n$ there must exist a fixed point.

In this direction one may also recall Newman's noteworthy result (1931) to the effect that the set of fixed points of a periodic homeomorphism of a manifold is nowhere dense. All these results also tie up with Hilbert's conjecture (Hilbert's fifth problem) that any locally Euclidean topological group is a Lie group.

5. Vector Fields.

The theory of vector fields has a surprising number of points of contact with topology. Consider first a vector field $\mathfrak{F}$ over a plane region Ω. A singular point of $\mathfrak{F}$ is a point where the vector is zero. The first problem of topological interest concerns the distribution of these points. Suppose that they are isolated and let P be one of them. Let γ be a small circuit surrounding P and no other singular point. As a point Q describes γ once in the positive direction of rotation, the angular variation of the vector $V(Q)$ of $\mathfrak{F}$ through Q is of the form $2m\pi$ where m is an integer independent of γ, and is called the index of P. There are many theorems regarding sums of indices of singular points. For example, if along a Jordan curve J not passing through any singular point, all the vectors point inwards or all outwards then the sum of the indices is unity.

One may associate with the field $\mathfrak{F}$ a deformation f of Ω such that the displacement of any point Q is proportional to the vector $V(Q)$ of $\mathfrak{F}$. The singular points of $\mathfrak{F}$ are then the fixed points of f. Instead of a plane, consider the same situation on a closed smooth surface (sphere, torus, etc.) or for that matter, on any compact differentiable manifold

M. Then the sum of the indices of the singular points (suitably defined by means of local coordinates) is equal to $\Phi(f) = \chi(M)$. As a consequence, for example for a spherical vector field $\chi(M) = 2$, hence there are always singular points. This is closely related to the nonexistence of a coordinate system valid over the whole sphere. All these results carry over to higher dimensions.

Of particular interest are the fields of the form grad$f(x_1, \cdots, x_n)$ (the vector has for components the partials $\partial f/\partial x_i$). Their singularities have been extensively investigated by Marston Morse. They correspond to the values z_i of z where the tangent space to $z = f(x_1, \cdots, x_n)$ is $z = z_i$. Thus for $n = 2$ there are generally just three types of contacts: maxima, minima, and saddle points. For any n the general singularities may be of any one of $n + 1$ types. Morse obtained a set of inequalities relating the numbers of the singularities of each kind (type numbers) to the Betti numbers of the domain provided that at the boundary the gradient always points inwards or outwards. Morse also applied the same results to the classification of types of extremals in the Calculus of Variation. This requires the study and calculation of Betti numbers of certain very complicated spaces of extremals. The topological situation encountered in this question is anything but simple.

Noteworthy vector fields arise also in the study of systems of linear differential equations of type

$$\frac{dx_i}{dt} = P_i(x_1, \cdots, x_n), \qquad (i = 1, 2, \cdots, n)$$

in the real domain. The dominating vector field $\mathfrak{F}$ is the field of vectors with the components P_i. Its singular points correspond in dynamical problems to positions of equilibrium. The closed orbits are also particularly important and the central problems are the determination of these orbits, their relation to the singular points and the stability of singular points and orbits. On all these questions there are basic contributions due to Poincaré, Liapounoff, Birkhoff and others.

The distribution of the orbits of systems

$$\frac{dx}{dt} = P(x,y), \qquad \frac{dy}{dt} = Q(x,y)$$

where P,Q are polynomials was extensively investigated by Poincaré (1886). The simplest singularities are of the types of Fig. 7 where in node and focus the arrows may also occur reversed. The respective indices are 1, -1, 1. On the other hand the sum of the indices of the singular points in a closed orbit γ is $+1$. For if along γ the vectors pointed along the interior normal the index would be $+1$. By revolving them by $\frac{\pi}{2}$ at each point they become tangent and the index is unchanged. Hence it is again $+1$. It follows that every closed orbit which surrounds only simple singular points must surround a node or a focus.

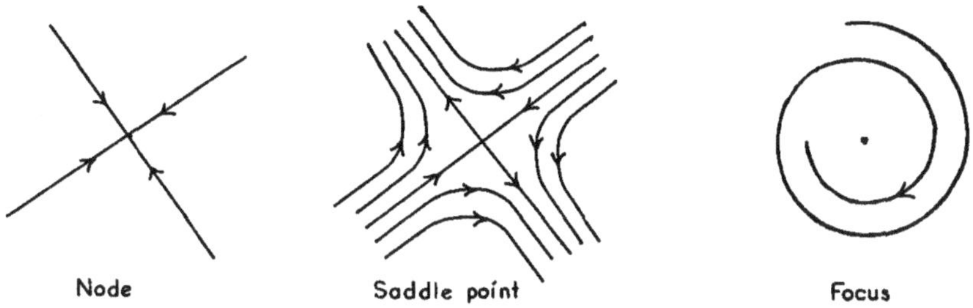

Figure 7

Another noteworthy topological result is the following theorem due to Poincaré and Bendixson: Let Ω be a bounded region with the following properties: (a) along its boundary B the vectors always point inwards or always outwards; (b) Ω is free from singular points. Then Ω contains a closed orbit. This is the only theorem of any degree of generality regarding the existence of closed orbits.

6. Integration and Topology.

It is not too much to say that both Riemann and Poincaré derived their inspiration for their topological contributions from the same basic problem: the study of the determinations of certain integrals with fixed limits of integration. It is from this direction in fact that algebra first penetrated into the citadel of topology.

Consider first a closed plane region Δ_p such as in Fig. 8. The boundaries consist of one large Jordan curve C and p smaller Jordan curves C_i. Consider now an integral of a total differential

$$J = \int U\,dx + V\,dy, \qquad \frac{\partial U}{\partial y} = \frac{\partial V}{\partial x}, \tag{7}$$

where U,V and their first partials are continuous in Δ_p. Let $J(\pi)$ denote the value of J taken along a path π. If π is a path of integration from a point A to a point B, for A fixed and B variable $J(\pi)$ becomes a function of B. What can be said regarding the possible determinations of this function? If π' is a second path from A to B and γ denotes the closed path from A to A consisting of π' followed by $-\pi$ (π reversed), then manifestly

$$J(\pi') = J(\pi) + J(\gamma). \tag{8}$$

Conversely if γ is any closed path from A to A, then γ followed by π is a π' such that (8) holds. The integral $J(\gamma)$ along a closed path γ is

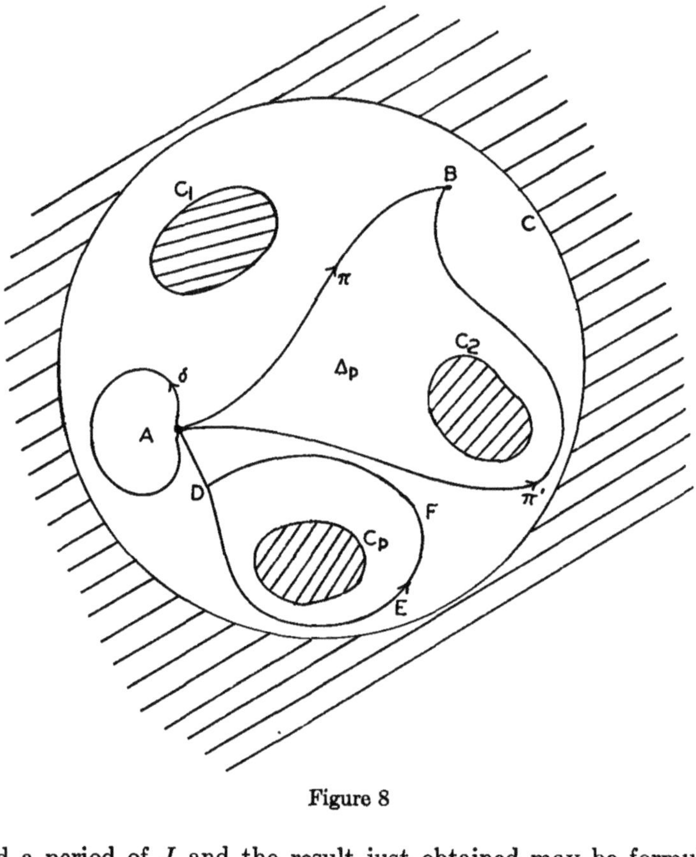

Figure 8

called a period of J and the result just obtained may be formulated thus: All the determinations of the integral of J from A to B are equal to a given determination $J(\pi)$ plus a period, and every such sum represents a determination.

The property just obtained merely shifts the problem of the determination of the integrals to the geometric investigation of the closed paths. From elementary analysis one learns that if a path such as δ in

Fig. 8 may be continuously deformed to a point $J(\delta) = 0$. Furthermore, a path or a portion of a path may be neglected if it consists of arcs described as many times in one direction as in the other. Thus in Fig. 8 the path $ADEFDA$ may be replaced by $DEFD$. If γ, γ' are reducible to one another in this manner, we will write a homology $\gamma \sim \gamma'$. The analogy with the homologies of Chapter III is obvious. Here again $\gamma \sim 0$ denotes "equal to something negligible." Let us also write: $-\gamma$ for γ reversed, $\gamma + \gamma'$ for γ followed by γ'. This gives content to a relation such as

$$(10) \qquad m\gamma \sim m_1\gamma_1 + \cdots + m_r\gamma_r,\ m_i \text{ an integer,}$$

and it implies

$$(11) \qquad m\,J(\gamma) = \Sigma m_i\,J(\gamma_i).$$

If the set $\{\gamma_i\}$ is such that: (a) a relation (10) exists for every γ; (b) no relation (10) exists with $m = 0$ and the m_i not all zero, then the set $\{\gamma_i\}$ is a "rational" base for the γ's. The number r of elements in the set is the one dimensional Betti number of Δ_p written $R^1(\Delta_p)$.

Under our assumptions the closed paths C_i (each oriented in a definite way) are admissible paths of integration. It may be shown that the set $\{C_i\}$ satisfies the two conditions (a), (b) just stated. Hence the set is a rational base and so $R^1(\Delta_p) = p$.

Of course (11) implies

$$(12) \qquad J(\gamma) = \Sigma\mu_i J(\gamma_i), \qquad \mu_i = m_i/m.$$

This suggests the formal extension of the scheme to homologies with rational coefficients μ_i:

$$(13) \qquad \gamma \sim \Sigma\mu_i\gamma_i,$$

as providing a more faithful description of the relations between the periods. It is also the reason for the term "rational base."

If we set $J(\gamma_i) = \omega_i$, the determinations of J are all of the form $J(\pi) + \Sigma\mu_i\omega_i, \mu_i$ rational. This type of expression is well known in analysis. It may be added that for the set $\{C_i\}$, the μ_i may all be chosen integers.

Noteworthy generalizations can easily be made here. The first is the possible replacement of Δ_p by a surface or a portion of surface with suitable related integrals. The classical instance is the study of abelian integrals where the stage for the development of algebraic topology may be said to have been set by Riemann's invention of the surface which bears his name. Let $f(x,y) = 0$ represent an irreducible algebraic curve Γ. An abelian integral attached to Γ is an integral of the form

$$\int R(x,y)dx, \tag{14}$$

where R is rational and y is the algebraic function $y(x)$ defined by $f = 0$. The integral is taken "in the complex domain" that is to say over paths made up of complex points of Γ. The totality of complex points of Γ may naturally be represented (in the sense of topology) as a "smooth" closed orientable surface (a compact orientable surface in the sense of Chapter II). This surface Φ_p with its family of analytic functions is the

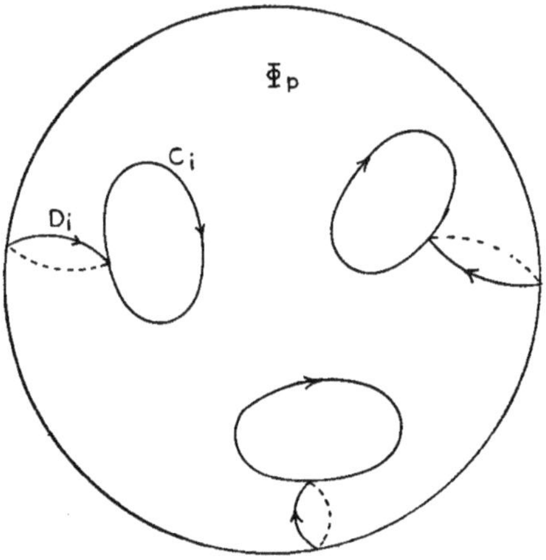

Figure 9

Riemann surface. As we shall show (see Chapter II) Φ_p is homeomorphic to a two-sided disk with p holes. The number p is the genus of Φ_p and it is readily computed (by a formula due to Riemann) in terms of simple algebraic properties of the polynomial f. It is thus on Φ_p that the paths of integration may now be drawn and compared.

The surface Φ_p may be constructed by fitting together two copies of the region of Fig. 8 along corresponding boundary curves.

To utilize the whole of the surface Φ_p as a region of integration one must have integrals (14) holomorphic everywhere on Φ_p. Such integrals are said to be of the first kind. The closed paths are shown to be reducible (in the homology sense) to the "retrosections" C_i, D_i of Fig. 9. Thus here $R^1(\Phi_p) = 2p$ and the periods of an integral of the first kind are linear rational combinations of $2p$ periods.

The simplest situation corresponds to the cubic curve $\Gamma : y^2 = 4x^3 - g_2x + g_3$. The associated Φ is a torus. There is an essentially unique integral of the first kind well known in the Weierstrass elliptic function theory:

$$\int \frac{dx}{y} = \int \frac{dx}{\sqrt{4x^3 - g_2x + g_3}} \tag{15}$$

and it has two periods.

Let us now consider multiple integrals. To avoid certain complications we shall assume that there is given a certain bounded closed region $\bar{\Omega}$ of ordinary space which is a polyhedron, i.e. decomposed into cells $\{E_i^p\}$, making up a complex K. The only continua of integration to be considered will be integral chains of K. Thus instead of the customary lines, surfaces, volumes of integration, we shall have one-chains, two-chains, three-chains. Except for this natural extension the properties described below are entirely familiar.

Let us suppose that $\bar{\Omega}$ is a vector field and let $V(A,B,C)$ be the general vector of the field. We will assume that A,B,C have continuous first partial derivatives throughout $\bar{\Omega}$. We recall the well known vector functions

$$\operatorname{div} V = \frac{\partial A}{\partial x} + \frac{\partial B}{\partial y} + \frac{\partial C}{\partial z} \tag{16}$$

$$\operatorname{curl} V = \left(\frac{\partial C}{\partial y} - \frac{\partial B}{\partial z}, \cdots\right). \tag{17}$$

The classical theorems are Green's theorem:

$$\int_{FE_i^2} V_t \, ds = \int_{E_i^2} \operatorname{curl} V \cdot d\sigma \tag{18}$$

and Stokes' theorem:

$$\int_{FE_i^3} V \, d\sigma = \int_{E_i^3} \operatorname{div} V \, d\tau, \tag{19}$$

where ds, $d\sigma$, $d\tau$ are the line, surface and volume differential elements and V_t the tangential component of V. By linear extension we obtain:

$$\int_{FC^2} V_t \, ds = \int_{C^2} \operatorname{curl} V \cdot d\sigma, \tag{20}$$

$$(21) \qquad \int_{FC^3} V\cdot d\sigma = \int_{C^3} \operatorname{div} V\, d\tau.$$

One may also consider integrals:

$$(22) \qquad \int_{C^1} V_t\, ds$$

$$(23) \qquad \int_{C^2} V\cdot d\sigma.$$

The n.a.s.c. in order that (22) depend solely upon the "limit" $FC^1 = \gamma^1$ is:

$$(24) \qquad \operatorname{curl} V = 0.$$

The n.a.s.c. in order that (23) depend solely upon the "limit" FC^2 is:

$$(25) \qquad \operatorname{div} V = 0.$$

These are more or less immediate and familiar consequences of the Green and Stokes relations. The role played by the properties just recalled in mathematical physics is only too well known. Thus if V is a force, (24) expresses the fact that the field is conservative.

The notion of differential form due to E. Cartan helps to clarify the situation. Let us set

$$(26) \qquad \varphi_1 = V_t\, ds = A\, dx + B\, dy + C\, dz,$$

$$(27) \qquad \varphi_2 = V_n\, d\sigma = A\, dy\, dz + B\, dz\, dx + C\, dx\, dy$$

$$(28) \qquad \varphi_3 = f(x,y,z)d\tau = f(x,y,z)\, dx\, dy\, dz$$

and call φ_p a p-form. Then $\operatorname{div} V\, d\tau$ is called the derived of φ_2, written $\delta\varphi_2$. Similarly $\operatorname{curl} V\cdot d\sigma$ is called the derived of φ_1, written $\delta\varphi_1$. We now have identically

$$(29) \qquad \delta\delta = 0.$$

If $\delta\varphi_p = 0$ we shall say that φ_p is exact. It is then easily seen that both Green and Stokes' relations assume the form

$$(30) \qquad \int_{C^{p+1}} \delta\varphi_p = \int_{FC^{p+1}} \varphi_p.$$

Green's theorem corresponds to $p = 1$ and Stokes' theorem to $p = 2$.

A consequence of (30) is that if γ^p is a cycle and φ_p is exact, the "period"

$$\omega(\gamma^p) = \int_{\gamma^p} \varphi_p \tag{31}$$

is the same for any two homologous cycles. The analogy with the case $p = 1$ and planar regions is thus complete.

It may be observed that relative to δ as boundary operator the forms behave like cochains with the derived forms as bounding cochains and exact forms as cocycles.

The whole argument may be extended to any orientable differentiable n-manifold M^n (i.e. with mutually differentiable local coordinate systems, the transformations from system to system having positive Jacobians throughout). Typical manifolds of this nature are ordinary smooth surfaces and the manifolds of relativity. Under suitable restriction de Rham proved (1931) the following noteworthy result: The number of φ_p's linearly independent modulo the derived forms is equal to the rational Betti number R^p of M^n. These results have been extended by W. V. D. Hodge to a more restricted class of forms and applied by him to algebraic geometry. See his recent monograph: *Harmonic Integrals* (1942). There are also other important connections with algebraic geometry for which we can only refer the reader to the author's monograph: *L'Analysis Situs et la Géométrie Algébrique* (Borel Series, 1923) and to Oscar Zariski's monograph, *Algebraic Surfaces* (1933).

I. Basic Information about Sets, Spaces, Vectors, Groups

The present chapter is merely intended as a summary of the basic definitions together with a statement of the principal theorems required in the following chapters. The unproved theorems, by far the largest part, are starred. The reader will find the proofs elsewhere in the literature but he is advised to work them out himself as the best method for absorbing this fundamental material.

§1. Questions of Notation and Terminology.

1. A moderate usage of the standard symbols and terms of the modern theory of sets is unavoidable. Those mainly occurring in the text are briefly recalled here.

We shall accept as known the common terminology of point-sets and in particular the terms: *set, collection, null set, function,* or *transformation.* We shall say "function f on A to B," or "transformation f of A *into* B," whenever for each point x of A the value fx is in B. This will sometimes be designated by $f : A \rightarrow B$. If A$'$ is a *subset* of A, the values of f on A' define a function f' on A' to B designated by $f|A'$ and f is known as the *extension* of f' to A.

With very rare exceptions functions are assumed *single-valued.* If f is a *one-one* transformation of A into fA, f is said to be *univalent.* If f is a transformation $A \rightarrow B$ and $fA = B$, the fact is sometimes underscored by saying "f transforms A *onto* B." If f transforms $A \rightarrow B$, and if C is a subset of B, we denote by $f^{-1}(C)$ the set of all points x of A such that fx is in C. The set $f^{-1}(C)$ is called the *inverse image* of the set C by f.

The statement: $\alpha \in A$, means that α is an *element* of the set A. If a_α, a_β, $\cdots$, are the elements of A, we shall often write $A = \{a_\alpha\}$, $A = \{a\}$, the meaning being usually clear from the context. If P is a property, the set of all elements of A which possess the property P is written $\{a|a \text{ has the property } P\}$. As an example, if $\{a\}$ is the set of all integers, the subset consisting of those divisible by 3 is written $\{a|a$ is divisible by $3\}$ or $\{a|a = 0 \bmod 3\}$.

If A and B are sets then $A \subset B$ or $B \supset A$ means: A is a subset of B. If $\{A_1, \cdots, A_r\}$, $\{A_\alpha\}$ are collections of sets, then $A_1 \cup \cdots \cup A_r$

or $\bigcup A_\alpha$ designates the *union* of the sets $A_1, \cdots, A_r$ or of the sets A_α, that is to say, the set whose elements are those belonging to any one of the sets $A_1, \cdots, A_r$ or to any one of the sets A_α. Similarly by $A_1 \cap \cdots \cap A_r$, $\bigcap A_\alpha$ is meant the *intersection* of the sets $A_1, \cdots, A_r$ or of the sets A_α, that is to say, the set whose elements are those belonging to every set $A_1, \cdots, A_r$ or to every set A_α. If A and B are sets, then $A - B$ denotes the set of all the elements of A which are not in B. It is also called the *complement* of B in A when $B \subset A$.

As an illustration of the preceding notations let A and B denote the sets of points interior to two intersecting circles as in Figs. 10, 11, 12.

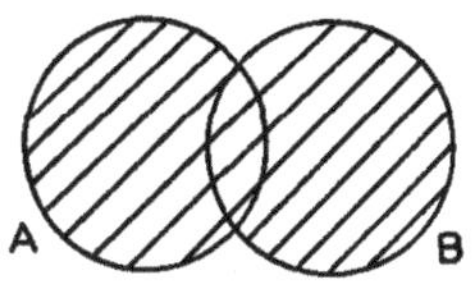

Figure 10

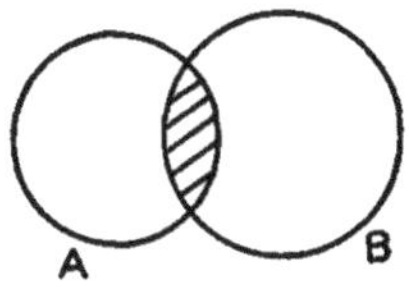

Figure 11

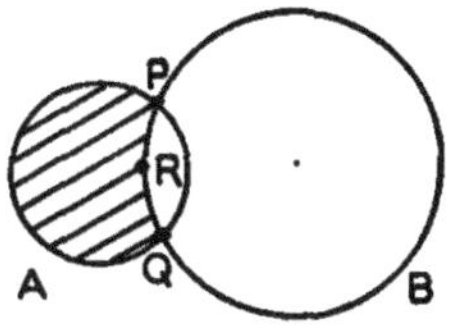

Figure 12

Then $A \cup B$ is the shaded area in Fig. 10, $A \cap B$ is the shaded area in Fig. 11, while $A - B$ is the shaded area in Fig. 12 plus the arc PRQ (end points P,Q excluded).

$A \cong B$ means: A is *isomorphic* with B (precise definition required each time).

The sets A_α are said to be *disjoint* whenever they do not intersect.

We shall also assume familiarity with the system of real numbers and all the usual related concepts, particularly those referring to convergence and limits. If the real line L is parameterized by means of a parameter u, then an *interval* is a set: $a < u < b$, and a *segment* is a set $a \leq u \leq b, a < b$. We shall also use on occasion the *supremum* and *infimum* of a nonvoid set A of real numbers, written sup A, inf A. Since they are not so well known, we shall say a few words about them. We may divide the set of all rational numbers into two classes: A_2 the set of all rational numbers which exceed every number of A, and A_1 those in A or else exceeded by some number of A. The class A_1 is certainly not empty. If neither A_1 nor A_2 is empty, then (A_1,A_2) is a *Dedekind cut* and the number which it determines is sup A. If A_2 is empty, we define $\sup A = +\infty$. Let B be the set of the negatives of the numbers of A. Then we define $\inf A = -\sup B$. Thus let A be the set of all rational numbers ≥ 1 and whose squares < 2. Then $\sup A = (A_1,A_2)$ is the real number $\sqrt{2}$, while $\inf A = -\sup B = 1$. Note that in this case sup A is not in A while inf A is in A.

*(1.1) **Theorem.** *If the set A has an upper [a lower] bound then* sup A [inf A] *exists.*

*(1.2) **Corollary.** *If $\{I_n\}$ is a collection of nested segments (closed intervals such that $I_{n+1} \subset I_n$) then there is a point common to all the segments (their intersection is nonempty).*

§2. EUCLIDEAN SPACES, METRIC SPACES, TOPOLOGICAL SPACES.

2. We shall deal mainly with *Euclidean sets:* Euclidean spaces or their subsets. The important facts regarding an n-dimensional Euclidean space $\mathfrak{E}^n$ are:

there is a one-one correspondence between the points of $\mathfrak{E}^n$ and a certain real coordinate system $(x_1, \cdots, x_n)$;

if the points x,y have the coordinates x_i,y_i, then there is a real function $d(x,y)$ called their *distance* (Euclidean distance), which serves as a basis for defining continuity, (see 3), given by

$$d(x,y) = \{\Sigma(x_i - y_i)^2\}^{1/2}. \tag{2.1}$$

The basic properties of this function, known as *distance axioms*, are:

D 1. $d(x,y) = 0$ when and only when $x = y$.

D 2. (*Triangle axiom.*) For all triples of points x,y,z:

$$d(y,z) \leq d(x,y) + d(x,z).$$

Two other important properties are:

D 3. $d(x,y) \geq 0.$

D 4. $d(x,y) = d(y,x).$

The only one not entirely obvious for the Euclidean distance (2.1) is the triangle axiom D 2. If we set $y_i - x_i = u_i$, $x_i - z_i = v_i$, then it reduces to:

$$\Sigma\,(u_i + v_i)^2 \leq \{(\Sigma u_i^2)^{1/2} + (\Sigma v_i^2)^{1/2}\}^2,$$

or to the inequality of Schwarz:

$$(\Sigma u_i v_i)^2 \leq (\Sigma u_i^2)(\Sigma v_i^2), \tag{2.2}$$

whose proof is as follows: If λ is any real number, then $\Sigma(u_i + \lambda v_i)^2 \geq 0$. If this is written in the form $A\lambda^2 + 2B\lambda + C$, then $B^2 - AC \leq 0$, and this last inequality written explicitly is precisely (2.2).

Metric spaces. A set of points $R = \{x\}$ with an attached real function $d(x,y)$ satisfying the distance axioms D 12 is known as a *metric space* and $d(x,y)$ is called the *distance-function* or merely *distance* for R. Thus an Euclidean space is a metric space whose distance function is the Euclidean distance (2.1). It may be shown that for any metric space R the properties D 34 are consequences of D 12. We prove D 3 by making $z = y$ in D 2. If we take $x = z$ in D 2 and remember D 1, we find $d(y,x) \leq d(x,y)$. Hence for obvious reasons of symmetry D 4 follows.

If $A \subset R$ and R is metric, $d(x,y)|A$ is a distance-function for A. Hence A is a metric space with this distance. It is this distance which will always be chosen henceforth for subsets of metric spaces.

A few basic concepts relative to metric spaces in general will now be recalled.

Distance between sets. Diameter of a set. Spheroids. If $A,B \subset R$, their distance written $d(A,B)$ is inf $\{d(x,y)|x \in A,\ y \in B\}$. The *diameter* of A, written diam A, is sup $\{d(x,y)|x,y \in A\}$. The *spheroid* of center x_o and radius ρ, written $\mathfrak{S}(x_o, \rho)$, is the set of all points x such that $d(x,x_o) < \rho$. In Euclidean geometry the spheroids are often called "*spherical regions.*"

Mesh; ϵ-aggregates; ϵ-transformations. If $A = \{A_\alpha\}$ is a collection of subsets of R, sup diam A_α is called the *mesh* of A, written mesh A. If mesh $A < \epsilon$, A is called an *ϵ-aggregate,* or *ϵ-collection.* If f transforms R into a set $Q = \{y\}$, and $\{f^{-1}y\}$, where $f^{-1}y$ is the total inverse image of y, is an ϵ-aggregate, then f is called an *ϵ-transformation.*

3. Questions of continuity. When dealing with transformations of metric spaces into one another, there is available the standard and well known "*ϵ,δ*" *principle of real variables* for deciding whether functions are continuous or not. We recall the exact statement. Let R,S be metric spaces and f a function on R to S. Then according to the principle just mentioned f is *continuous* whenever if $y = fx$, and if $\epsilon > 0$ and x are given, there exists a corresponding $\delta > 0$ such that if $d(x,x') < \delta$ and $y' = fx'$ then $d(y,y') < \epsilon$.

Topological transformations. Topology. If f is one to one and *bicontinuous* (both f and its inverse f^{-1} continuous), f is said to be a *topological transformation* or a *homeomorphism. Topology* may now be described as the study of the properties of configurations which are unaffected by topological transformations.

It is readily seen that the usual properties of real continuous functions in real variables carry over to continuous functions on metric spaces to metric spaces.

Cells, disks, spheres. These sets, as simple as they are important, will recur constantly in the sequel. Consider the *Euclidean sphere* in $\mathfrak{E}^n$:

$$\Sigma x_i^2 = 1.$$

An *open n-cell,* or merely *n-cell* is any homeomorph of the set $\Sigma x_i^2 < 1$ (interior of the sphere). A *topological* $(n - 1)$-*sphere,* or merely $(n - 1)$-*sphere,* is any homeomorph of the sphere itself. A zero-sphere consists of two points. A one-sphere, the homeomorph of a circumference, is also known as a *Jordan curve.* A zero-cell is a point. A homeomorph of the set $\Sigma x_i^2 \leq 1$ is variously called: *closed n-cell, solid n-sphere,* also *n-disk* (A. W. Tucker). An interval is a one-cell, a segment is a one-disk.

*(3.1) *An Euclidean n-space* $\mathfrak{E}^n$ *is an open n-cell.*

4. Topological spaces. Metric spaces are so useful and convenient that it is natural to search for other possible spaces with similar structural properties. To define them one must have recourse to some inner properties of suitable classes of sets which may be the open sets or the closed sets. We first examine the nature of these sets in metric spaces, and by analogy define the new spaces so as to preserve their principal properties.

Open sets. A subset U of a metric space R is said to be an *open set* of R whenever U is the null set or else if $x \in U$, then there is a spheroid $\mathfrak{S}(x,\rho) \subset U$. Incidentally this makes U a union of spheroids.

(If R is a Euclidean space the term *region* is often used instead of "open set.")

*(4.1) *Every spheroid is an open set. Hence any union of spheroids is an open set.*

*(4.2) *The open sets of a metric space satisfy the following three axioms:*

OS 1. *The null set and the space itself are open sets.*

OS 2. *The union of any number of open sets is an open set.*

OS 3. *The intersection of any two (hence of any finite number) of open sets is an open set.*

Given any point set $\mathfrak{R} = \{x\}$, we may set up as its system of open sets any system of subsets $\{U\}$ satisfying the system of axioms OS 123. A set $\mathfrak{R}$ to which there is thus assigned a collection of open sets satisfying OS 123 is said to be *topologized* or to have received a *topology* and $\mathfrak{R}$ is then called a *topological space.*

It may be observed that in a metric space the following axiom correlating points and open sets is likewise fulfilled:

OS 4. *Given any two distinct points* x,y *there exist disjoint open sets* U_x, U_y *containing respectively* x,y.

A topological space satisfying OS 4 is known as a *Hausdorff* space. Thus a metric space is a Hausdorff space.

Notice that any point set $\mathfrak{R} = \{x\}$ always contains a collection of subsets turning it into a Hausdorff space, namely the collection of *all* the subsets of $\mathfrak{R}$. In particular each point of $\mathfrak{R}$ is an open set in this topology. The special topology thus arising is known as *discrete,* and $\mathfrak{R}$ topologized in this manner is known as a *discrete* topological space.

Given a topological space $\mathfrak{R}$ there may exist a distance-function whose spheroids give rise to the open sets of $\mathfrak{R}$. When this is the case $\mathfrak{R}$ is said to be *metrizable.* It is hardly necessary to observe that the metric when it exists is certainly not unique. For if $d(x,y)$ is a suitable distance function so are $kd(x,y)$, $k > 0$ and $\dfrac{d(x,y)}{1 + d(x,y)}$.

If two metrics both give rise to the same open sets for a set $\mathfrak{R}$, they are said to be *equivalent.* Thus the distance-functions for $\mathfrak{E}^n$ given by

$$\{\Sigma(x_i - y_i)^2\}^{1/2}, \qquad \Sigma|x_i - y_i|$$

define equivalent metrics.

A *base* for a topological space $\mathfrak{R}$ is a collection $\{W\}$ of nonvoid open sets such that every nonvoid open set of $\mathfrak{R}$ is a union of sets W.

Example. The Euclidean space $\mathfrak{E}^n$ referred to the coordinates $x_1, \cdots, x_n$ possesses the *countable base* consisting of all the sets $|x_i - a_i| < b_i$ where a_i and b_i are rational and $b_i > 0$.

A convenient way to topologize a set R is to specify that a certain collection $\{W_a\}$ is to be a base, that is to say that the open sets are to consist of the null set and of all the unions of sets of the collection. We then have:

*(4.3) *N.a.s.c. for $\{W_a\}$ to be admissible as a base are:* (a) *R is a union of sets W_a;* (b) *the intersection of any two sets W_a is a union of sets W_a.*

*(4.4) *If $A \subset \mathfrak{R}$ and $\{W\}$ is a base for $\mathfrak{R}$ then $\{A \cap W\}$ is a base for A. Hence if $\mathfrak{R}$ has a countable base so have all its subsets. In particular every Euclidean set has a countable base.*

A space $\mathfrak{R}$ is said to be *separable* whenever it contains a countable subset $\{x_n\}$ which is *dense* in $\mathfrak{R}$ (every open set contains some x_n).

*(4.5) *For a metric space separability and the possession of a countable base are equivalent properties.*

*(4.6) **Theorem of Lindelöf.** *If $\mathfrak{R}$ has a countable base and $V = \bigcup V_a$ where the V_a are open sets in $\mathfrak{R}$ then there exists a countable subcollection $\{V_{a_n}\}$ of $\{V_a\}$ such that $V = \bigcup V_{a_n}$.*

*(4.7) **Corollary.** *If $\mathfrak{R}$ has a countable base and $\{V\}$ is any base*

whatever then there exists a countable subcollection $\{V_{a_n}\}$ *of* $\{V_a\}$ *which is already a base.*

*(4.8) *The spheroids* $\mathfrak{S}(x_p,\rho_q)$ *of* $\mathfrak{E}^n$ *where* $\{x_p\}$ *are all the rational points and* ρ_q *are all the rational radii, form a (countable) base for* $\mathfrak{E}^n$.

5. Further properties of topological spaces. *Closed sets.* The complement $F = \mathfrak{R} - U$, where U is an open set, is said to be a *closed set.*

*CS 1. *The null set and* $\mathfrak{R}$ *itself are closed sets.*

*CS 2. *The intersection of any collection of closed sets is closed.*

*CS 3. *The union of any two (and hence of any finite number of) closed sets is closed.*

Conversely, if we started with the closed sets $\{F\}$ as collections satisfying CS 123, then defined the open sets U as their complements, the collection $\{U\}$ would satisfy OS 123. Thus there is a large measure of symmetry between the two concepts.

Closure. Given any set A in $\mathfrak{R}$, we define its *closure* $\bar{A}$ as the intersection of all the closed sets containing A. It may be thought of as the "least" closed set containing A. Clearly $A = \bar{A}$ when and only when A is closed.

Interior. The *interior* of a set A is defined to be the union of all the open sets contained in A. Thus the interior of a set is the largest open set contained in it. An *interior point* of a set is a point belonging to the interior of the set.

Neighborhood. A *neighborhood* of a set A is any open set containing A. Thus if $\mathfrak{R}$ is metric $\mathfrak{S}(x,\rho)$ is a neighborhood of x.

The closure $\bar{A}$ of A is the set of all points x such that every neighborhood of x intersects A.

A set A is dense in $\mathfrak{R}$ if $\bar{A} = \mathfrak{R}$ or equivalently if A meets every open set of $\mathfrak{R}$.

*(5.1) *If* $\mathfrak{R}$ *is metric,* $\bar{A}$ *is the set of all points of* $\mathfrak{R}$ *whose distance from* A *is zero.*

Boundary of a set. The *boundary* of A, written $\mathfrak{B}A$, is by definition the set $\bar{A} \cap \overline{\mathfrak{R} - A}$. If $\mathfrak{R}$ is metric $\mathfrak{B}A$ is the set of all points at zero distance from both A and its complement.

Example. In the Euclidean plane Π let C be a circumference, U the region interior to C, λ an arc of C. Set $A = U \cup \lambda$. Then $\bar{A} = \bar{U} = U \cup C$, and $\mathfrak{B}A = \mathfrak{B}U = C$.

Principle of relativization. If $\mathfrak{R}$ is a topological space and A is a subset of $\mathfrak{R}$ then the collection $\{A \cap U\}$ of the intersections with A of the open sets U of $\mathfrak{R}$ satisfies the axioms OS 123. We are therefore justified in agreeing once for all that A is made a topological space by adopting for its open sets the collection $\{A \cap U\}$. The closed sets of A are then the intersections $A \cap F$ of A with the closed sets of $\mathfrak{R}$. The sets $A \cap U$, $A \cap F$ are sometimes said to be *relatively open, relatively closed* in A. Hence the name "*principle of relativization*" generally given to the scheme just adopted.

Notice that if $\mathfrak{R}$ is metric and $A \subset \mathfrak{R}$ then the open sets of A defined by means of the distance on A are precisely the intersections with A of those of $\mathfrak{R}$. In other words the principles of relativization for the distance and for the open sets agree.

6. Continuity for topological spaces. There exist topological spaces for which it is impossible to define a distance which would essentially reproduce the defining open sets of the space. However in the absence of a distance one may still define continuity in various ways. The procedure adopted here is by means of the open sets. Let f be a transformation $\mathfrak{R} \to \mathfrak{S}$, where $\mathfrak{R}$, $\mathfrak{S}$ are topological spaces. If then for every open set V in $\mathfrak{S}$, the set $U = f^{-1}(V)$ is open in $\mathfrak{R}$, the function f is said to be *continuous*. Explicitly f is continuous if the inverse images of the open sets of $\mathfrak{S}$ under f are open sets in $\mathfrak{R}$.

Let f be one-one, so that in particular $f\mathfrak{R} = \mathfrak{S}$. Then f is said to be a *topological transformation* of $\mathfrak{R}$ *onto* $\mathfrak{S}$ when both f and f^{-1} are continuous. This means that every fU is a set V and every $f^{-1}V$ is a set U.

Topology for topological spaces is now defined as before.

Thus for metric spaces there appear to be two types of continuity.

*(6.1) *The two types of continuity are the same. That is to say they specify the same functions as continuous.*

Hereafter it will not be necessary to distinguish between the two kinds of continuity in metric spaces, and we shall merely say "continuous function, continuous transformation." These transformations are also referred to as *mappings*.

7. Connected sets. A topological space A is said to be *connected* if we cannot write $A = B \cup C$, where B and C are nonempty disjoint open subsets (hence also closed subsets) of A. A subset A of a topological space $\mathfrak{R}$ is therefore connected, by the principle of relativization, if we do not have $A = B \cup C$, where B,C are nonempty, disjoint and open (hence also closed) in A. It is clear from the definition that connectedness is a topological property.

*(7.1) *A union of connected sets of which every pair intersect is itself connected.*

*(7.2) *A sequence $\{A_n\}$ of connected sets such that each meets the next one has a connected union.*

*(7.3) *If a set A is such that any two points x,y of A are contained in a connected subset B of A then A is connected.*

If x is any point of A, the union $C(x)$ of all the connected subsets of A containing x is a connected set, the *component* of A determined by x. If $y \in C(x)$, then each of $C(x)$, $C(y)$ contains the other and so they coincide. Thus a component is determined by any one of its points. For this reason we refer to $C(x)$ merely as a component of A. If A has a single component then it is connected and conversely.

(7.4) *Application. The open n-cell, the closed n-cell, likewise for $n > 0$ the n-sphere, are all connected.*

We first show that a segment $l : 0 \leq t \leq 1$ is connected. Suppose that the contrary is true. Then $l = A \cup B$, where A,B are nonempty, closed and disjoint in l. Let $0 \in A$ and set $a = \sup \{t|t \in A, t < B\}$. Since A is closed, $a \in A$. In addition whatever $\eta > 0$ there is a point of B in the interval a, $a + \eta$. Hence $a \in B$, since B is closed. Since this contradicts the assumption that A and B are disjoint, l is connected.

Since any two points of one of the sets in the statement may be joined by a *closed arc* (topological image of l), they are all connected.

*(7.5) *The closure of a connected set is connected. Hence a component is a closed set.*

*(7.6) *The image of a connected set under a mapping is connected.*

*(7.7) *If the number ρ of components of a set A is finite A is the union of ρ disjoint closed sets. If ρ is infinite then whatever r the set A is the union of r disjoint closed sets.*

§3. COMPACT SPACES.

8. The property of compactness may be introduced from different points of view. Our approach will be by means of the open coverings, its advantage being that it is suitable for any topological space, in fact for any space which has any sets labelled "open."

Let us recall a few definitions. A *covering* of a space $\mathfrak{R}$ is a collection $\mathfrak{A} = \{A_\alpha\}$ of subsets whose union is $\mathfrak{R}$: $\bigcup A_\alpha = \mathfrak{R}$. If $\mathfrak{R}$ is metric and every diam $A_\alpha \leq \epsilon$, we also call $\mathfrak{A}$ an *ϵ-covering*. If all the A_α are open sets [closed sets] then $\mathfrak{A}$ is said to be *open* [*closed*.] If $\mathfrak{A}$,$\mathfrak{B}$ are two coverings and every set of $\mathfrak{A}$ is contained in some set of $\mathfrak{B}$, then $\mathfrak{A}$ is called a *refinement* of $\mathfrak{B}$, and said to *refine* $\mathfrak{B}$.

We will say that a collection of sets $\{A_\alpha\}$ has the *finite intersection property* if every finite subcollection has a nonempty intersection.

Consider now the following two possible properties of a topological space $\mathfrak{R}$:

P 1. *If $\{U_\alpha\}$ is any open covering then some finite subcollection of $\{U_\alpha\}$ is already a covering.*

P 2. *If* $\{F_\alpha\}$ *is a collection of closed sets with the finite intersection property then the intersection of the whole collection is nonempty.*

It is not difficult to show that the two properties are equivalent. Suppose first P 1 true, and let $\{F_\alpha\}$ be as in P 2 and yet $\bigcap F_\alpha$ empty. If we set $U_\alpha = \mathfrak{R} - F_\alpha$, the last property implies that $\{U_\alpha\}$ is an open covering. By assumption P 1 holds and hence some finite subcollection $\{U_{\alpha_i}\}$ is a covering. This implies however that $\bigcap F_{\alpha_i}$ is empty, which contradicts the hypothesis that $\{F_\alpha\}$ has the finite intersection property. Therefore $\bigcap F_\alpha$ is not empty and so P 1 implies P 2.

Suppose now that P 2 holds and let $\{U_\alpha\}$ be an open covering. If $F_\alpha = \mathfrak{R} - U_\alpha$ then $\bigcap F_\alpha$ is empty, and so by P 2 some finite subcollection $\{F_{\alpha_i}\}$ has an empty intersection. This implies however that $\{U_{\alpha_i}\}$ is a covering, or P 1 holds. Thus P 2 implies P 1, and so P 1, P 2 are equivalent.

A topological space possessing one of the properties P 12 (and therefore both) is said to be *compact*. Notice that by virtue of the principle of relativization for a subset A of a topological space $\mathfrak{R}$, the condition P 1 may be replaced by: every covering $\{U_\alpha\}$ of A by open sets of $\mathfrak{R}$ (i.e. the union of the U_α contains A) has a finite subcovering.

The following are noteworthy and also convenient properties of compact topological spaces.

*(8.1) *A closed subset F of a compact topological space $\mathfrak{R}$ is also compact.*

*(8.2) *Let $\mathfrak{R}$ be compact and f a mapping of $\mathfrak{R}$ into a topological space $\mathfrak{S}$. Then the image $f\mathfrak{R}$ is also compact.*

*(8.3) *The union of a finite number of closed subsets of a space is compact if and only if each closed subset is compact.*

*(8.4) *The image of a compact space $\mathfrak{S}$ under a mapping f into a Hausdorff space $\mathfrak{R}$ is closed in $\mathfrak{R}$. Hence if f is one-one it is topological and both $\mathfrak{R}$ and $\mathfrak{S}$ are compact Hausdorff spaces.*

9. Compacta. A compact metrizable space is called a *compactum*. The class of compacta and their subsets includes by far the most interesting sets occurring in topology. Thus closed cells, spheres, polyhedra, projective spaces are all compacta.

*(9.1) *For a metric space compactness is equivalent to the following condition: every infinite subset has a point of condensation.*

(A *point of condensation* of a set $E \subset \mathfrak{R}$ is a point x such that every neighborhood of x contains points of $E - x$.)

(9.2) **Theorem.** *A n.a.s.c. for a Euclidean set A to be compact (and hence to be a compactum) is that it be closed and bounded.*

Suppose A compact in $\mathfrak{E}^n$ and take a fixed point $x \in \mathfrak{E}^n - A$. The collection $\{A \cap \mathfrak{S}(y, \frac{1}{2}\, d(x,y)) | y \in A\}$ is an open covering of A. Since A is compact there is a finite subcovering. That is to say there is a finite subset $\{y_i\}$ of A such that if $d(x,y_i) = 2\rho_i$ then $A \subset \bigcup \mathfrak{S}(y_i, \rho_i)$. Hence diam $A < 2 \Sigma \rho_i$ is finite, or A is bounded. Moreover if $\rho = \inf \rho_i$ then $\mathfrak{S}(x,\rho) \subset \mathfrak{E}^n - A$. Hence $\mathfrak{E}^n - A = \bigcup \mathfrak{S}(x,\rho)$ is open and so A is closed. Thus the condition of the theorem is necessary.

Suppose now A closed and bounded in $\mathfrak{E}^n$. Then A is contained in some *parallelotope* $\Pi : |x_i| \leq b$, and since Π is closed, A is closed in Π. Thus to show that A is compact it is sufficient to prove:

(a) Π *is compact.*

Since $\mathfrak{E}^n$ is separable so is Π. Hence by Lindelöf's theorem every open covering of Π has a countable subcovering. Thus the argument reduces to showing that:

(b) *Every countable open covering* $\{U_m\}$ *of* Π *has a finite subcovering.*

The proof follows a well-known pattern. Consider the successive finite subcollections $\mathfrak{U}_m = \{U_1, \cdots, U_m\}$. If (b) is incorrect no $\mathfrak{U}_m$ is a covering, and so for every m there is an x_m which is in no set U_p, $p \leq m$. We thus obtain an infinite sequence $\{x_m\}$. Consider the partition $P_q = \{\pi_i^q\}$ of Π into 2^{qn} parts obtained by dividing each edge into 2^q equal parts. Since the partition P_1 is finite and m takes an infinite number of values there must be a set π_1' of P_1 containing an infinite subsequence $\{x_{m'}\}$. Repeating the reasoning for the subsequence and the partition of π_1' by P_2, we obtain a similar $\{\pi_2'\}$, etc. We thus have a sequence $\{\pi_r'\}$ such that: (a) every π_r' contains an infinite number of points of $\{x_m\}$; (b) $\pi_{r+1}' \subset \pi_r'$. The projections of the π_r' on any axis make up a sequence of nested segments whose diameters $\rightarrow 0$ and so by (1.2) the segments intersect in a point. Hence $\bigcap \pi_r'$ is a point x of Π. This point has manifestly the property that every $\mathfrak{S}(x,\rho)$ hence every neighborhood of x, contains points of $\{x_m\}$ of arbitrarily high index m, that is to say, x is a condensation-point of the sequence.

Since $\{U_m\}$ is a covering, x is in some U_r and there is an x_m of arbitrarily high index m contained in U_r, and therefore in a set of $\mathfrak{U}_m$. Since this contradicts the choice of $\{x_m\}$, (b) follows. This proves the sufficiency of the condition and completes the proof of the theorem itself.

10. Application to closed cells, and spheres. It is an immediate consequence of the preceding results that:

(10.1) *Closed cells and spheres are compacta.*

Of more practical value is a general property of convex sets which will enable us to disclose many simple sets as closed cells or spheres.

We recall that by definition a *convex set* is a subset A of an Euclidean space such that if x,y are in A then so is the segment xy which joins them. We will now prove:

(10.2) *A bounded closed convex region Ω of $\mathfrak{E}^n$ is a closed n-cell.*

Let B denote the boundary of Ω and let P be any interior point of $\Omega(P \in \Omega - B)$ and S a sphere of center P. Denote by H the closed spherical region bounded by S. Any ray issued from P meets B in a single point Q and S in a single point R. Let t be the transformation $\Omega \rightarrow H$ defined as follows. If M is any point of the segment PQ then

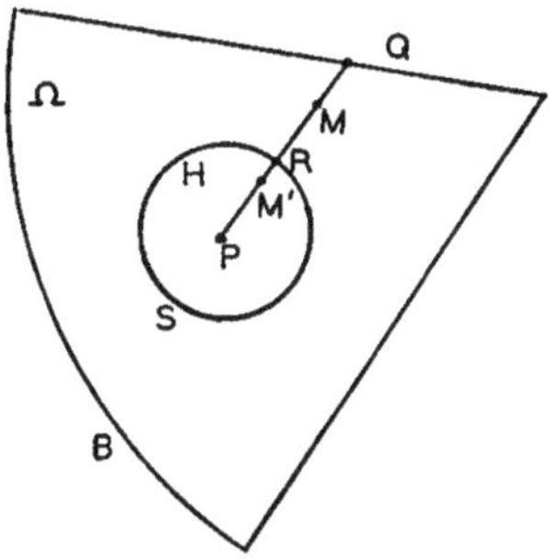

Figure 13

$M' = tM$ is that point of the segment PR which divides it in the same ratio as M divides PQ. It is easily seen that t is one-one and continuous. Hence since H is Hausdorff and Ω is compact, t is topological (8.4) and since H is a closed n-cell, this proves (10.2).

*(10.3) *Let $\mathfrak{E}^p \subset \mathfrak{E}^n$, and let Ω be a closed convex region of $\mathfrak{E}^p$ (= closure of a relatively open convex subset). Then Ω is a closed p-cell.*

11. Further properties of compacta. A few additional properties are recalled. In the statements $\mathfrak{R}$ is a compactum, $\mathfrak{F} = \{F_1, \cdots, F_r\}$ a finite collection of closed sets, $\mathfrak{U} = \{U_1, \cdots, U_s\}$ a finite collection of open sets.

*(11.1) *If the sets of $\mathfrak{F}$ do not intersect, there is a positive number $c(\mathfrak{F})$ such that every point x of $\mathfrak{R}$ is at a distance $\geq c(\mathfrak{F})$ from some F_i.*

*(11.2) *Corresponding to any $\mathfrak{F}$ there is a positive number $d(\mathfrak{F})$, called the Lebesgue number of $\mathfrak{F}$, such that if $A \subset \mathfrak{R}$ and* diam *$A < d(\mathfrak{F})$ and A meets a collection of sets of $\mathfrak{F}$ then the sets of the collection intersect.*

*(11.3) *If $\mathfrak{U}$ is a covering there is a corresponding positive number $d_1(\mathfrak{U})$, called the Lebesgue number of $\mathfrak{U}$, such that:* (a) *every point x of $\mathfrak{R}$ is in some U_i and at a distance $> d_1(\mathfrak{U})$ from its complement $\mathfrak{R} - U_i$;* (b) *if $A \subset \mathfrak{R}$ and* diam *$A < d_1(\mathfrak{U})$, then A is contained in some set U_i.*

§4. VECTOR SPACES.

12. Due to their simplicity these spaces will be extensively utilized in connection with homology. However, before taking them up we shall say a few words regarding groups and fields.

Groups. We shall merely mention that unless otherwise stated all groups are assumed *additive*, i.e. commutative, with addition as the group operation, and unity written 0. A *subgroup* G' of a group G is a subset of G such that if g_1', g_2' are in G' so is their difference $g_1' - g_2'$. If G, H are two groups, then a *homomorphism* $\tau : G \to H$ is a transformation such that $\tau g - \tau g' = \tau(g - g')$. The inverse $G' = \tau^{-1}0$ is a subgroup of G called the *kernel* of τ. If τ is one-one it is known as an *isomorphism* of G with H.

Let G be a group and H a subgroup. By a *coset* of G relative to H (also called a coset of G mod H) we understand a subset H' of G, such that there exists a fixed $g \in G$ such that $H' = \{g + h | h \in H\}$. Two cosets are clearly either disjoint or coincident. The cosets of G relative to H from a group called the *factor-group* of G mod H, written G/H. The transformation $\pi : G \to G/H$ sending each element of G into its coset is a homomorphism, called the *projection* of G into G/H, and its kernel is H.

We shall have repeated occasion to deal with the following question: Given two groups G,H and a homomorphism $\tau : G \to H$, to prove that τ is an isomorphism. This will always be done by reference to one of the following two criteria:

*(12.1) *If τ is univalent and onto then it is an isomorphism.*

*(12.2) *If there exist homomorphisms $\theta,\theta' : H \to G$ such that $\theta\tau = 1$, $\tau\theta' = 1$ then τ is an isomorphism and $\theta = \theta' = \tau^{-1}$.*

It may be observed that both criteria are also valid for *multiplicative* (i.e. perhaps noncommutative) *groups.*

Fields. We recall that a *field* is merely a collection of elements $\Omega = \{\alpha\}$ subject to the same rules of rational operations as ordinary (real or complex) numbers. Examples of fields are: (a) the set of all complex numbers; (b) the set of all real numbers; (c) the set of all rational numbers; (d) the set of all (integral) residues modulo a prime number π. This last field is finite and consists of π elements. The two fields (c) and (d) will be particularly important in dealing with the homology groups.

13. Vector spaces. A *vector space* V over a field Ω is an additive group $\{x\}$ such that there is defined a product αx for every pair (α,x), $\alpha \in \Omega$, $x \in V$, with values in V, and with the properties:

$$1x = x,\ 0x = 0,\ \alpha(\alpha' x) = (\alpha\alpha')x,\ (\alpha + \alpha')x = \alpha x + \alpha' x,$$
$$\alpha(x + x') = \alpha x + \alpha x'.$$

When Ω is the real or the rational field or the set of residues mod π, V is said to be *real* or *rational* or *mod* π.

Example. Let $\xi_1, \cdots, \xi_n$ be n arbitrary symbols and define as a vector any linear form $x = \alpha_1\xi_1 + \cdots + \alpha_n\xi_n$, $\alpha_i \in \Omega$ with addition in the obvious way and the conventions

$$1\xi_i = \xi_i,\ 0\xi_i = 0,\ \alpha x = \Sigma(\alpha\alpha_i)\xi_i.$$

Then $\{x\}$ is a vector space over Ω. When Ω is the field of the residues mod π the vector space just defined is finite and consists of π^n elements.

Let $V = \{x\}$ be a vector space over Ω. The elements $x_1, \cdots, x_r$ of V are said to be *linearly dependent* if there exists a relation $\Sigma\alpha_i x_i = 0$, with the $\alpha_i \in \Omega$ not all zero; in the contrary case the x_i are said to be *linearly independent.* If there is a finite maximum n of linearly independent elements we say that n is the *dimension* of V, written dim V. When no finite maximum exists we merely say that the dimension of V is infinite. The field Ω itself is a one-dimensional vector space.

Suppose $n = \dim V$ is finite and let $\xi_1, \cdots, \xi_n$ be independent elements. If x is any other element whatsoever we have a relation

$$\alpha x + \alpha_1\xi_1 + \cdots + \alpha_n\xi_n = 0$$

with the $\alpha, \alpha_1, \cdots, \alpha_n \in \Omega$ not all zero. We cannot have $\alpha = 0$ since then the ξ_i would be linearly dependent. Thus $\alpha \neq 0$ and so if we set $-\beta_i = \frac{\alpha_i}{\alpha}$, we have

$$x = \beta_1\xi_1 + \cdots + \beta_n\xi_n,\ \beta_i \in \Omega. \tag{13.1}$$

Moreover, this representation of x is unique. For suppose

$$x = \gamma_1\xi_1 + \cdots + \gamma_n\xi_n,\ \gamma_i \in \Omega,$$

where the $\beta_i - \gamma_i = \delta_i$ are not all zero. Since $\Sigma\delta_i\xi_i = 0$ the ξ_i would be linearly dependent. Thus every element x is represented in one and only one way in the form (13.1) and all expressions (13.1) represent elements of V. For this reason $\{\xi_i\}$ is called a *base* for V.

Example. In the first example of the present number the $\{\xi_i\}$ form a base and so the dimension is n.

Linear transformations. Isomorphism. If V,W are vector spaces over Ω and if φ is a (group) homomorphism $V \to W$, then φ is called a *linear transformation* of V into W if $\varphi\alpha x = \alpha\varphi x$. The term *isomorphism* as between V,W is reserved for group isomorphisms which are likewise linear transformations. The explicit characteristics

of a linear transformation φ are: $\varphi(\alpha x + \beta u) = \alpha\varphi x + \beta\varphi y$,

of $V \cong W$ are: φ is one-one and $\varphi(\alpha x + \beta y) = \alpha\varphi x + \beta\varphi y$.

Since the zeros correspond in an isomorphism the relations of linear dependence are preserved. Hence:

*(13.2) *If $V \cong W$ and one of the two spaces is finite dimensional so is the other and the dimensions are the same.*

*(13.3) *Conversely if V,W have equal finite dimensions then $V \cong W$. Thus for finite dimensional vector spaces over the same field Ω, equality of the dimensions is a n.a.s.c. for isomorphism.*

Subspaces. A subset W of V which is likewise a vector space over Ω is known as a *subspace* of V. In the following statements we suppose that W is a subspace of V and that $\dim V = n$, $\dim W = m$.

*(13.4) $\dim W \leq \dim V$ *and the two are equal when and only when* $W = V$.

*(13.5) *If W is a subspace of V, there exists a base $\{\xi_1, \cdots, \xi_m, \eta_1, \cdots, \eta_{n-m}\}$ for V such that $\{\xi_i\}$ is a base for W.*

*(13.6) *Under the same circumstances let U be the subspace spanned by the η_i* (i.e. *subspace of the vectors $\Sigma\alpha_i\eta_i$*). *Then every $v \in V$ may be written uniquely in the form $v = u + w$, $u \in U$, $w \in W$. The space V is then called the direct sum of U and W, written $V = U + W$. We also have* $\dim U = \dim V - \dim W$ *and U is isomorphic with the factor-group V/W. The space U is generally not unique.*

§5. PRODUCTS OF SETS, SPACES AND GROUPS. HOMOTOPY.

14. Products. The *product* of a finite collection of sets $X = \{x\}, \cdots, Y = \{y\}$ is merely the *set* of all collections $(x, \cdots, y)$ of one element from each set, and is written $X \times \cdots \times Y$. When the sets are topological spaces with open sets $\{U\}, \cdots, \{V\}$, the *product* is made a topological space, and sometimes called *topological product* of $X, \cdots, Y$, by choosing as base for its open sets the collection $\{U \times \cdots \times V\}$. The verification of the condition (4.3) is immediate and so the choice of base just made is legitimate. Whenever $X, \cdots, Y$ are groups, their product is made into a group by defining $0 = (0, \cdots, 0)$, $(x, \cdots, y) + (x', \cdots, y') = (x + x', \cdots, y + y')$. Finally if $X, \cdots, Y$ are vector spaces over Ω, the product is made a vector space over Ω by the convention $\alpha(x, \cdots, y) = (\alpha x, \cdots, \alpha y)$, $\alpha \in \Omega$.

Examples. The topological product of two real lines, X, Y referred to the coordinates x, y is a cartesian plane whose base consists of all the horizontal or vertical strips and open rectangles parallel to the axes. The topology is easily shown to be the same as that of the Euclidean plane. Similarly the topological product of n real lines is a cartesian n-space homeomorphic with $\mathfrak{E}^n$.

The topological product of n intervals is an open n-cell, and the topological product of n segments is a closed n-cell.

The product of n copies of a field Ω is an n-dimensional vector space over Ω.

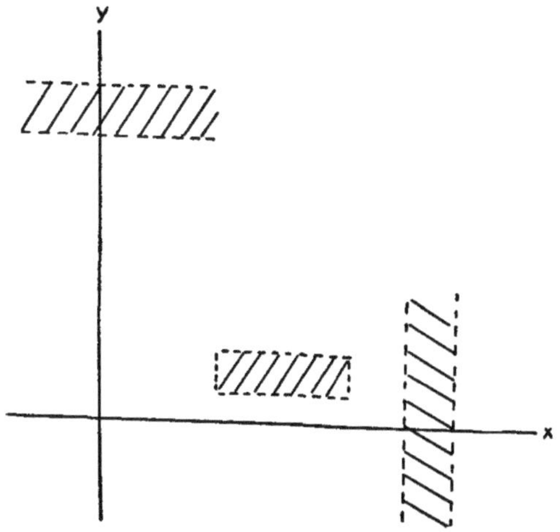

Figure 14

15. Homotopy, deformation. The importance of these concepts in modern topology could scarcely be exaggerated. The topological product will greatly facilitate their description.

Let A, B be topological spaces and consider the totality $\{f\}$ of all mappings of A into B. In some sense one realizes that whenever it is possible to pass "continuously" from a mapping f_1 to a mapping f_2, they are to be considered as in the same "family." What is needed is to give a more precise content to the words in quotes. To illustrate what we have in mind let A be a circumference and B a torus with α,α',β as in Fig. 15. They are the images of A under mappings f_α, $f_{\alpha'}$, f_β. It is intuitively apparent that one can pass continuously from f_α to $f_{\alpha'}$ but not from f_α to f_β.

The situation is completely cleared up by means of the concept of *homotopy*. Let l denote the segment $0 \leq u \leq 1$ and consider the *cylin-*

der $l \times A$. The two mappings f,g of A into B are said to be *homotopic*, written $f \backsim g$, whenever there exists a mapping F of the cylinder $l \times A$ into B which agrees with f on $0 \times A$ (the first base) and with g on $1 \times A$ (the second base). That is to say if $x \in A$ then $F(0 \times x) = fx$ and $F(1 \times x) = gx$.

The image of the generator $l \times x$ under F is called the *path* of x in the homotopy. If B is metric and diam $F(l \times x) < \epsilon$ for all $x \in A$, we say that we have an *ϵ-homotopy*. When f is the identity, which implies that $A \subset B$, the homotopy is said to be a *deformation*.

It is not difficult to see that in Fig. 15 there is a deformation of α into α', the paths being arcs of parallels (if we suppose that each parallel

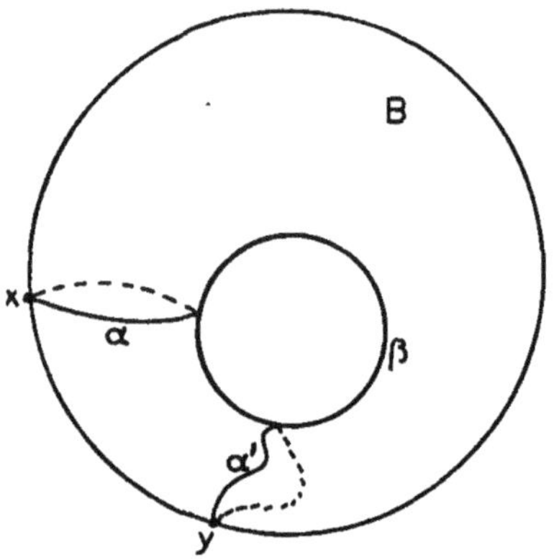

Figure 15

meets α or α' in a single point). One of the paths would be for example the arc xy in Fig. 15.

If $f \backsim g$ such that gA is a point f is said to be *homotopic to a constant*. If the identity in A is homotopic to a constant A is said to be *deformable into a point*, also to be *contractible*.

*(15.1) *Homotopy is an equivalence relation.*

*(15.2) *If* $f \backsim g$ *(as mappings* $A \to B$*) and* φ *is a mapping* $B \to C$ *then* $\varphi f \backsim \varphi g$ *(as mappings* $A \to C$*).*

*(15.3) *If B is contractible every mapping into B is homotopic to a constant.*

*(15.4) *If* $B \subset \mathfrak{E}^n$ *and for each* $x \in A$, fx *and* gx *may be joined by a segment in B then* $f \backsim g$. *Similarly for* $B = S^n$ *and segments replaced by arcs of great circles through a point P not in fA nor gA, where the arcs* $\subset S^n - P$.

*(15.5) *If A is a compactum so is the cylinder $l \times A$ and hence likewise its image $F(l \times A)$ in B if B is metrizable.*

Mappings into an Euclidean space $\mathfrak{E}^n$ or a sphere S^n are particularly interesting. Such a mapping f into one or the other is said to be *inessential* if it is homotopic to a constant and to be *essential* otherwise.

*(15.6) *Every mapping of a compactum into an Euclidean space is inessential.*

*(15.7) *If A is a compactum and f is a mapping $A \to S^n$ such that fA is a proper subset of S^n then f is inessential.*

*(15.8) *If $\bar{E}^{n+1}$ is a closed cell and S^n is its boundary sphere then a n.a.s.c. for a mapping $f: S^n \to A$ to be homotopic to a constant is that it may be extended to a mapping $f_1: \bar{E}^{n+1} \to A$.*

Let $\mathfrak{E}^n$ be a subspace of $\mathfrak{E}^{n+1}$, P a point of $\mathfrak{E}^{n+1} - \mathfrak{E}^n$ and A a subset of $\mathfrak{E}^n$. The set Γ of all segments joining P to the points of A is known as a *cone* of base A and vertex P.

*(15.9) *Let f be a mapping of the compactum A into a topological space $\mathfrak{R}$. A sufficient condition in order that f be homotopic to a constant is that it may be extended to a mapping of the cone Γ into $\mathfrak{R}$.*

PROBLEMS

1. Obtain the necessary and sufficient conditions to be satisfied by the points and open sets in order that all the points of a topological space $\mathfrak{R}$ be closed. A space in which this holds is said to be a *T_1 space.* Construct an example of a T_1 space which is not a Hausdorff space.

2. If A is compact then a continuous real valued function f on A assumes its supremum and infimum.

3. Let $X = \{x(t)\}$ be the set of all real continuous functions on $0 \leq t \leq 1$. Show that the following functions are distance functions:

$$\sup|x(t) - y(t)|, \left[\int_0^1 (x-y)^2\, dt\right]^{1/2}, \int_0^1 |x - y|\, dt.$$

4. If A is a subset of a metric space $\mathfrak{R}$ let $\mathfrak{S}(A,r) = \bigcup \{\mathfrak{S}(x,r) | x \in A\}$. Define now the distance of two closed subsets A,B as

$$d(A,B) = \inf \{r | A \subset \mathfrak{S}(B,r),\, B \subset \mathfrak{S}(A,r)\}.$$

Prove that under this definition of distance the closed subsets of $\mathfrak{R}$ form a metric space $\mathfrak{F}$. Prove that if $\mathfrak{R}$ is compact so is $\mathfrak{F}$.

5. Let $x = \{x_1, x_2, \cdots\}$, where the x_i are non-negative real numbers and Σx_i^2 converges. Show that the set $X = \{x\}$ is a metric space with a distance

$$d(x,y) = [\Sigma(x_i - y_i)^2]^{1/2}.$$

This space is called *Hilbert space*. Show that it contains all the points x such that for every $n : 0 \leq x_n \leq 1/n$. Show that the set of all these points is compact. It is known as the *Hilbert parallelotope*.

6. Let t be a transformation of a metric space $\mathfrak{R}$ into a metric space $\mathfrak{S}$. A n.a.s.c. for t to be continuous is that if $\{x_n\}$ is a convergent sequence in $\mathfrak{R}$ and x its limit, then $\{tx_n\}$ is a convergent sequence in $\mathfrak{S}$ with the limit tx.

7. A topological product of Hausdorff spaces is a Hausdorff space.

8. If A and B are compact so is $A \times B$.

9. A compact subset of a Hausdorff space is closed.

10. A component of M is closed in M.

11. If C and M are connected and C is contained in M and A is a component of $M - C$, then $M - A$ is connected.

12. Every compactum is separable.

13. Let X, Y be two topological spaces and $\mathfrak{F} = \{f\}$ the collection of all mappings $X \to Y$. Given a compact subset A of X and an open set B of Y let $(A,B) = \{f | f A \subset B\}$. The finite intersections of the sets (A,B) may be taken as a base and $\mathfrak{F}$ with the resulting topology is denoted by Y^X and called a *function space*. Prove the following properties:

(a) If Y is a Hausdorff space, so is Y^X.

(b) If X is a compactum and Y has a countable base, then Y^X has a countable base.

(c) If X is compact and Y is metric, then Y^X can be metrized by the formula

$$\xi(f,g) = \sup \{d(f(x),g(x)) | x \in X\}.$$

(d) If X consists of n points, then Y^X is homeomorphic with Y^n.

(e) If X is compact, Y metric and contractible, then Y^X is connected.

(f) If X is a segment, then X^X is noncompact.

II. Two-Dimensional Polyhedral Topology

The main topics dealt with in the present chapter—complexes of dimension at most two, the Jordan curve theorem, classification of surfaces—were primarily selected and treated so as to provide a suitable preparation for what follows. This has made it possible to introduce very early and in their simpler form a number of the basic methods of the general theory. Regarding the Jordan curve theorem, the proof given here, while not as simple as some, does lend itself so easily to extensions to higher dimensions that these have been relegated to the exercises.

§1. Elements of the Theory of Complexes. Geometric Considerations.

1. As already stated only complexes of dimension at most two are to be considered in this chapter. They arise quite naturally out of the decomposition of certain figures into triangles, arcs and points. The systematic development of the theory requires however that one proceed in the opposite direction by putting together the constituent elements, the *simplexes*, like a house built from bricks.

We shall say: *n-simplex, n-complex*, $\cdots$, for n-dimensional simplex, n-dimensional complex, $\cdots$, and wherever needed exhibit the dimension as a superscript: σ^n, $\cdots$.

A zero-simplex σ^0, also called a *vertex*, is merely a point. A one-simplex σ^1 is an arc (open one-cell) which has two distinct end-points not considered, however, as part of σ^1. Thus the closure $\bar{\sigma}^1$ is a closed one-cell. Since it is the topological image of a segment, $\bar{\sigma}^1$ may be parametrized by a real variable whose range is $0 \leq u \leq 1$. This means in particular that in proving any topological statement about $\bar{\sigma}^1$ we may replace it by the segment.

A two-simplex σ^2 is an open rectilinear or curvilinear triangle (perimeter excluded). Thus a spherical triangle without its perimeter is a σ^2. More explicitly σ^2 is an open two-cell whose closure $\bar{\sigma}^2$ is a closed two-cell whose boundary (the set $\bar{\sigma}^2 - \sigma^2$) is made up of three vertices a,b,c and three disjoint arcs ab, bc, ca, the vertices and arcs forming together a Jordan curve. Thus $\bar{\sigma}^2$ may be represented topologically as a closed circular region whose boundary circumference is decomposed into three arcs by three subdivision points. Here again in proving topological properties of $\bar{\sigma}^2$ one may substitute for it the closed circular region with its boundary subdivided in the manner just described.

The zero-simplex σ^0 is its own unique *face*. The simplex σ^1 and its

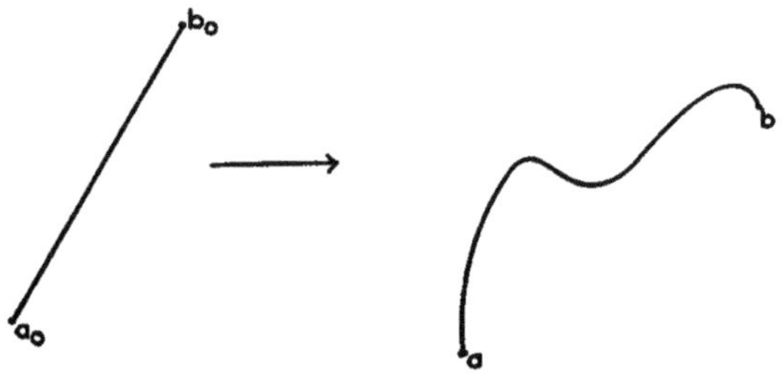

Figure 16

end-points are the faces of σ^1. The simplex σ^2 and the three arcs and vertices in its boundary are the faces of σ^2. A *proper face* of σ^s is a face other than σ^s itself. Thus σ^0 has no proper face, σ^1 has two, σ^2 has six. A simplex and its face are said to be *incident with one another*.

The simplexes will now be put together to form *complexes*. Under the restrictions of the present chapter (the dimensions must not exceed

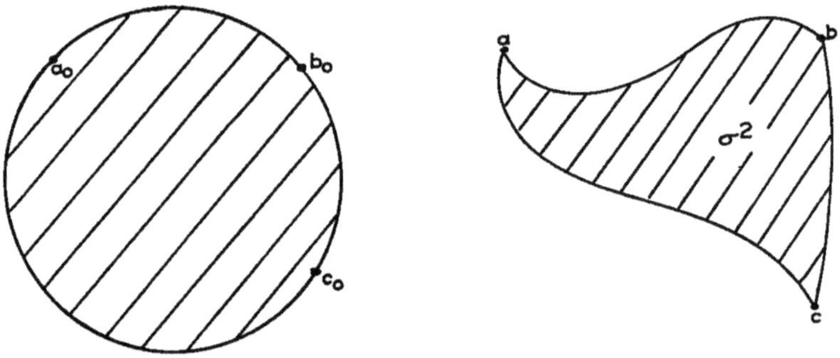

Figure 17

two) a complex is a finite collection $K = \{\sigma_i^0, \sigma_j^1, \sigma_k^2\}$ of vertices, arcs and triangles satisfying the following two conditions:

I. The simplexes of K are disjoint and no two have the same vertices.

II. If a simplex is in K all its faces are likewise in K.

Thus if σ^1 is in K its end-points are vertices of K, while if σ^2 is in K its sides and vertices are simplexes of K. It is important to bear in mind the negative statements: no vertex is in an arc or a triangle, no arc ends in a triangle or meets a triangle.

Simple instances of complexes are: the collection of all the faces, or of all the proper faces, of a simplex, a triangulated sphere or plane region.

The point-set union of all the simplexes of a complex K, or for that matter of any collection K of simplexes, whether a complex or not, is written $|K|$. When K is a complex, $|K|$ is called a *polyhedron*. The polyhedron is said to be *covered* by K. We also refer sometime to K as a *triangulation* or *subdivision* of $|K|$. A given polyhedron Π, for instance the sphere, may be *triangulated* in various ways. A standard method

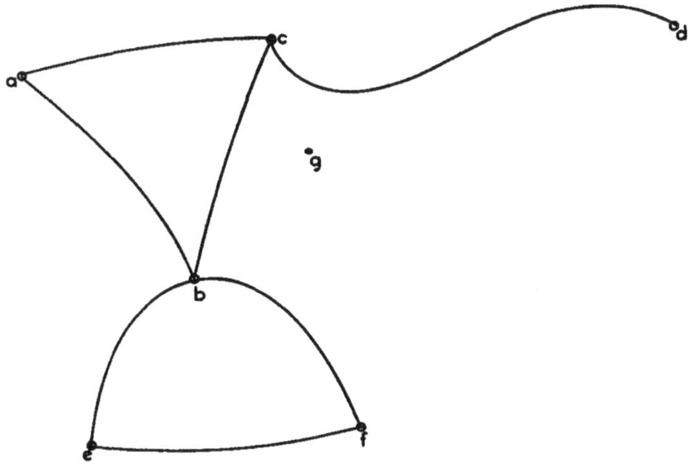

Figure 18

of topology consists in looking for properties common to all the covering complexes of Π, as they are necessarily topological properties of the polyhedron.

The *dimension* n of K is the maximum dimension of its simplexes and K is also called an n-complex. If $\dim K = 1$, K is also called a *linear graph* or merely a *graph*. The faces, edges and vertices of a tetrahedron or of a triangulated sphere make up a two-complex. The arcs and vertices resulting from the subdivision of a Jordan curve by a finite number of points make up a graph.

A *subcomplex* L of K is a subcollection of its simplexes which is a complex.

It is often convenient to name the vertices $a,b, \cdots$, and then to designate the simplexes as ab, cd, $\cdots$, abc, $\cdots$. This will cause no confusion since under our conventions distinct simplexes of K never have the same vertices.

2. Two complexes $K = \{\sigma\}$, $K_1 = \{\zeta\}$ are said to be *isomorphic* if there exists between their elements a one-one correspondence $\sigma_i^s \leftrightarrow \zeta_i^s$ which preserves the incidences: a vertex or side of σ_i^s corresponds to a vertex or side of ζ_i^s and conversely.

(2.1) *When K, K_1 are isomorphic there is a homeomorphism $t : |K| \to |K_1|$ such that $t\sigma = \zeta$ where σ is the element corresponding to ζ under the isomorphism.*

Let $a_1, \cdots, a_r$ be the vertices of K and $b_1, \cdots, b_r$ those of K_1, the numbering being such that a_i,b_i correspond under the isomorphism. Define now $ta_i = b_i$. If $\sigma^1 = a_ia_j$, its image in K_1 under the isomorphism is $\zeta^1 = b_ib_j$. Let $\bar{\sigma}^1,\bar{\zeta}^1$ be parametrized respectively by $0 \leq u \leq 1$, $0 \leq v \leq 1$, with a_i,a_j as the points $u = 0$, $u = 1$, and b_i,b_j as the points $v = 0$, $v = 1$. Then t has the obvious topological extension obtained by choosing as the image of u of $\bar{\sigma}^1$ the point $v = u$ of $\bar{\zeta}^1$.

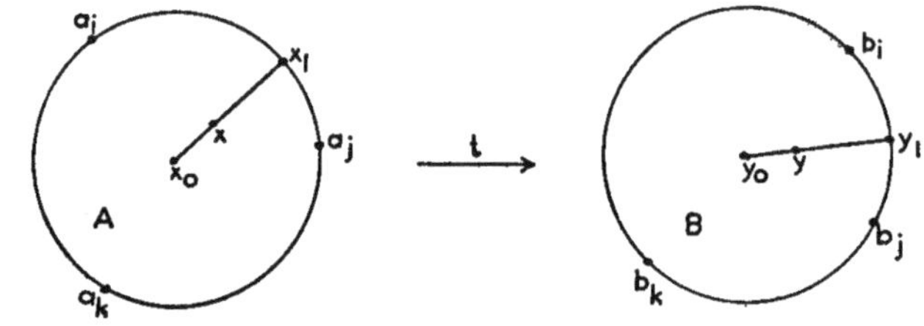

Figure 19

Suppose now t thus extended to all the σ_i^0, σ_j^1 and let $\sigma^2 = a_ia_ja_k \in K$ and $\zeta^2 = b_ib_jb_k$ its image in K_1 under the isomorphism. Choose as topological images of $\bar{\sigma}^2,\bar{\zeta}^2$ two closed circular regions A,B of equal radii and of centers x_0,y_0. If x_1 is in the boundary of A then $tx_1 = y_1$ is known and is in the boundary of B. If x is in the segment $x_0\bar{x}_1$ define $y = tx$ as the point of the segment $y_0\bar{y}_1$ such that $d(x_0,x) = d(y_0,y)$. Evidently t is thus topologically extended to σ^2.

If the arcs of K are rectilinear intervals and its triangles are ordinary plane triangles, all in some fixed Euclidean space $\mathfrak{E}^r$, K *is* called a *geometric complex.* Thus the triangles, edges and vertices of a tetrahedron make up a geometric complex.

Let $K = \{\sigma\}$ be any complex and $K_1 = \{\zeta\}$ a geometric complex which is an isomorph of K in the same manner as above, with an attached homeomorphism $t : |K| \to |K_1|$ such that $t\sigma = \zeta$. We call K_1 a *geometric antecedent* of K.

(2.2) *Every complex has a geometric antecedent.*

Let again $a_1, \cdots, a_r$ be the vertices of K. In a Euclidean r-space $\mathfrak{E}^r$ with coordinates $x_1, \cdots, x_r$ let b_i be the point whose coordinates are δ_{ij} ($x_i = 1, x_j = 0$ for $j \neq i$). For each $\sigma = a_i \cdots a_j \in K$ construct the geometric simplex (vertex, interval or triangle) $\zeta = b_i \cdots b_j$. Evidently $K_1 = \{\zeta\}$ is a geometric isomorph of K. By (2.1) a suitable homeomorphism $|K| \to |K_1|$ exists and it will make K_1 a geometric antecedent of K.

The existence of a geometric antecedent K_1 of K will enable us, as regards all topological questions, to replace K by K_1, i.e. to assume K geometric. We shall then say briefly that K is taken in a *geometric representation.*

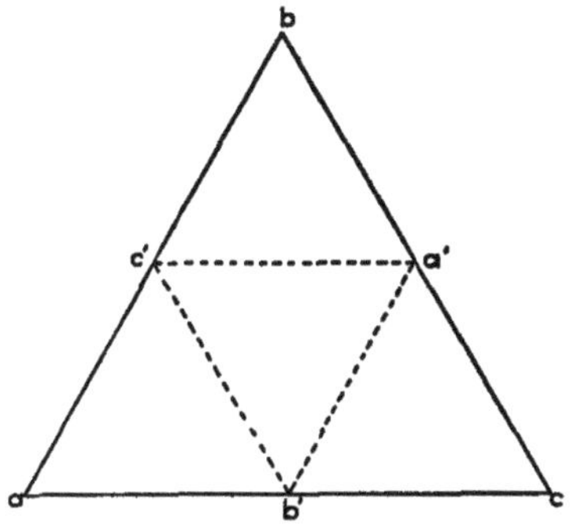

Figure 20

3. Subdivision. Among the various types of subdivision that may be defined we shall choose one particularly suited to our general purpose.

Given K let us take it in a geometric representation and define as an *elementary subdivision* a complex K' obtained from K by bisecting all the one-simplexes and subjecting all the triangles to the operation illustrated in Fig. 20. The vertices of K' are those of K and in addition the midpoints $a', \cdots,$ of the sides of K. Thus ab is replaced by c', ac', $c'b$ and abc by the four small triangles and those of their sides and vertices which are *not* in the perimeter of abc. A *subdivision* K_1 of K is obtained by repeating the preceding operation a finite number of times. If $K = \{\sigma\}$ is any complex and $K_1 = \{\zeta\}$ a geometric isomorph with t as the mapping $|K| \to |K_1|$ of (2.1), then a subdivision L of K is obtained as follows: take a subdivision L_1 of K_1; the images of the simplexes of L_1 under t^{-1} make up L.

(3.1) *Every complex K has a subdivision K_1 whose mesh is arbitrarily small.*

Let L be a geometric antecedent of K. Since $|L|$ and $|K|$ are homeomorphic and compact a subdivision L_1 of L is imaged into one K_1 of K and mesh $K_1 \to 0$ with mesh L_1. Hence we may replace K by L in the proof of (3.1), and therefore we may choose K in a geometric representation. Let D be the operation of elementary subdivision in K. It is evident that D halves the diameters of the simplexes of K, and hence it halves the mesh of K. Therefore D^n divides the mesh by 2^n from which (3.1) follows.

§2. ELEMENTS OF THE THEORY OF COMPLEXES. MODULO TWO THEORY.

4. Chains and cycles. A fundamental role will be assigned in the sequel to certain linear forms in the simplexes σ_i^s with integral coefficients reduced mod 2. The coefficients may therefore be assumed at any time to be 0 or 1, and it will be possible to dispense with negatives.

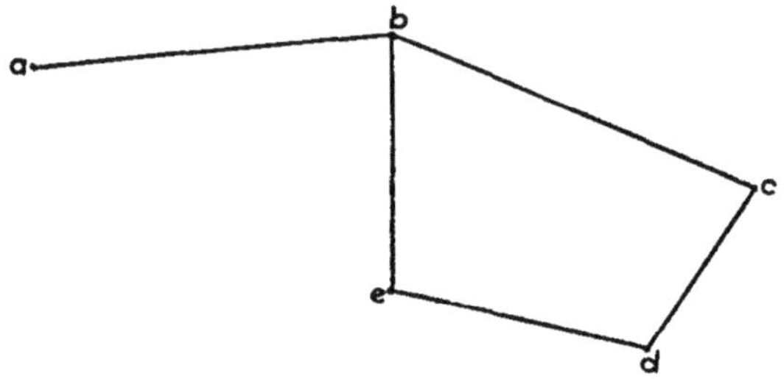

Figure 21

(4.1) Let $K = \{\sigma\}$ be a complex. An *s-chain* of K, or in K, is merely a form $C^s = \Sigma g_i \sigma_i^s$. We write $C^s = 0$ if every g_i is even. If $C'^s = \Sigma g'_i \sigma_i^s$ is a second chain of K then by definition $C^s + C'^s = \Sigma(g_i + g'_i)\sigma_i^s$. The *simplex* σ_i^s is said to be *in* C^s, written $\sigma_i^s \subset C^s$, if g_i is odd. The compactum which is the union of the point sets $\bar{\sigma}_i^s$, $\sigma_i^s \subset C^s$ is denoted by $|C^s|$. Fig. 21 illustrates a one-chain $C^1 = ab + bc + cd + de + eb$.

With each C^s we associate a new chain FC^s, its *chain-boundary*, defined as follows: $FC^0 = 0$; if $s > 0$ and $C^s = \Sigma g_i \sigma_i^s$, then $FC^s = C^{s-1} = \Sigma h_j \sigma_j^{s-1}$, where h_j is the number of σ_i^s in C^s which are incident with σ_j^{s-1}. Thus F is linear: $F(C^s + C'^s) = FC^s + FC'^s$. In particular,

$$Fa = 0, \qquad Fab = a + b, \qquad Fabc = bc + ca + ab.$$

From this follows identically $FF\sigma^s = 0$, and since FF is linear every $FFC^s = 0$. In Fig. 21 $FC^1 = a + b$.

An *s-cycle* of K is a chain C^s such that $FC^s = 0$. Thus in Fig. 21 $bc + cd + de + eb$ is a one-cycle but C^1 is not. Since F is linear the sum of two s-cycles is an s-cycle. Also every C^0 is a cycle. Finally since

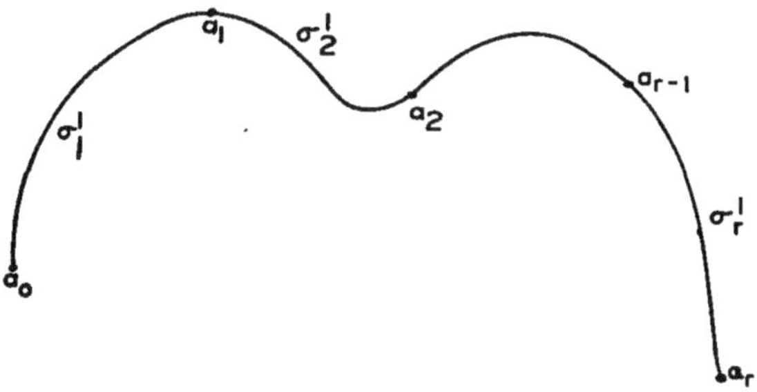

Figure 22

$F(FC^s) = 0$, $C^{s-1} = FC^s$ is a cycle. Such a cycle is said to *bound*, written $C^{s-1} \sim 0$. We also write $C^s \sim C'^s$ if $C^s + C'^s \sim 0$. The relation $\sim$ is called a *homology* and it is evidently an equivalence.

It is worth while to keep in mind, at least at first, certain simple analogies which help to guide intuition. Thus a chain and its boundary

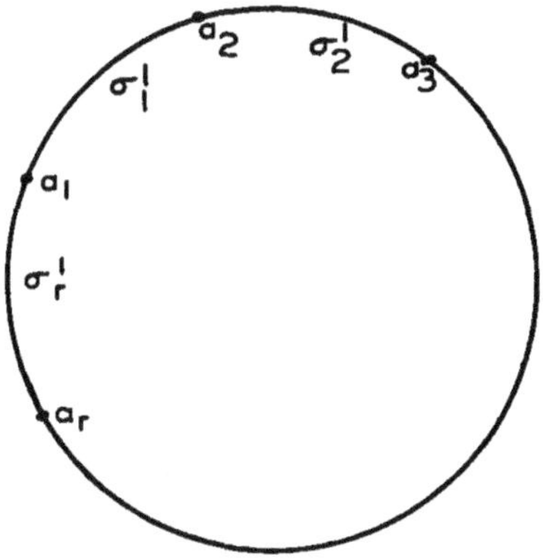

Figure 23

are suggested by a polyhedron and its boundary (in the sense of boundary or border of a hemisphere). The cycle is suggested by a closed polyhedron (such as the sphere). The property for a cycle to be ~ 0 suggests its being negligible because it is collapsible into a cycle with all simplexes taken an even number of times (cycle zero).

Some examples. (4.2) *The subdivided arc.* This is a complex K obtained from subdivision of an arc. In the notations of Fig. 22 if $C^1 = \Sigma g_i \sigma_i^1$ is to be a one-cycle and contains σ_i^1, $i < r$, then it must contain σ_{i+1}^1, else FC^1 would contain a_i. Thus a one-cycle must be of the form $C^1 = \sum_{i=s}^{r} \sigma_i^1$, $s \leq r$. But $FC^1 = a_{s-1} + a_r \neq 0$, since $a_{s-1} \neq a_r$. Hence the subdivided arc has no one-cycle $\neq 0$.

(4.3) *The subdivided circumference.* This time the situation is that of Fig. 23 with $r \geq 3$. Make the convention that $a_{r+i} = a_i$. Then as before the only possible one-cycle $\neq 0$ is $\gamma^1 = \Sigma\sigma_i^1$ and since $F\gamma^1 = 0$, γ^1 is a cycle. Thus the subdivided circumference has a single one-cycle $\gamma^1 \neq 0$ and it is the sum of its one-simplexes.

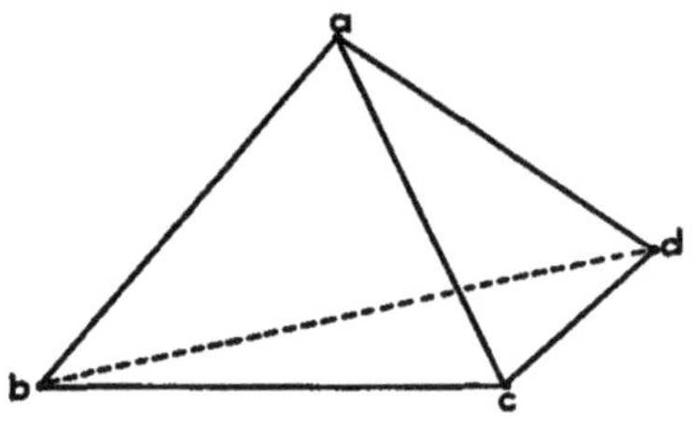

Figure 24

(4.4) *The tetrahedron boundary.* Let K denote the complex made up of the triangles, edges, vertices of the tetrahedron $abcd$. It is easily seen that

(4.5) $\gamma^2 = \Sigma\sigma_i^2$ *is a two-cycle of* K *and it is the only one* $\neq 0$.

Less evident is the property

(4.6) *Every one-cycle of* K *bounds.*

If a one-cycle γ^1 contains the simplex ab, it must have exactly one more with the vertex a, i.e. it must have one of the simplexes ac or ad. Hence γ^1 is of one of the types

$$ab + bc + ca = Fabc \sim 0,$$

$$ab + bc + cd + da = F(abc + adc) \sim 0.$$

This proves (4.6).

5. Chain-mapping. This is an algebraic analogue and constant companion of transformations and partitions of polyhedra—hence its fundamental importance.

Given two complexes K, K_1, suppose that there exists a linear transformation φ on the chains of K to those of K_1 of equal dimension: φC^s

is an s-chain; $\varphi(C^s + C'^s) = \varphi C^s + \varphi C'^s$. We refer briefly to φ as a *linear transformation* $K \to K_1$. Owing to linearity φ is fully defined by its values on the simplexes σ of K. We say that φ is a *chain-mapping* whenever it commutes with the boundary operator $F : \varphi F C^s = F\varphi C^s$ for all C^s of K. This is also written in operator form $\varphi F = F\varphi$. Since F and φ are both linear the commutation condition is equivalent to $\varphi F\sigma = F\varphi\sigma$ for every $\sigma \in K$.

The following two properties are immediate:

*(5.1) *A chain mapping φ sends C^s and FC^s respectively into a chain φC^s and its boundary $F\varphi C^s$. Hence φ sends a cycle into a cycle and a bounding cycle into a bounding cycle.*

*(5.2) *If K, K_1, K_2 are complexes and φ, ψ are chain-mappings $K \to K_1$, $K_1 \to K_2$, then $\psi\varphi$ is a chain-mapping $K \to K_2$.*

6. Chain-subdivision. Side by side with the geometric operation of subdivision there is an algebraic companion linear operation d known as *chain-subdivision*. If $K_1 = \{\zeta\}$ is a subdivision of $K = \{\sigma\}$ then d is a linear transformation on the chains of K to those of K_1 such that $d\sigma^s$ is the sum of the s-simplexes of K_1 contained in σ^s. Thus in Fig. 20: $d(a) = a$, $d(ab) = ac' + c'b$, $d(abc) =$ the sum of the four smaller triangles of the figure.

The definition yields at once:

(6.1) *dC^s is the sum of the s-simplexes of K_1 contained in those of C^s.*

(6.2) $|dC^s| = |C^s|$.

(6.3) *If L is a subcomplex of K and $C^s \subset L$, then dC^s consists of s-simplexes of K_1 contained in those of L.*

Of more importance is property

(6.4) *d is a chain-mapping.*

Owing to the product property of chain-mappings it is sufficient to prove (6.4) when K_1 is an elementary subdivision. Then since d is linear it is sufficient to prove $dF\sigma^s = Fd\sigma^s$ for any $\sigma^s \in K$. For a σ^0 this is trivial. For $\sigma^1 = ab$ and $\sigma^2 = abc$ we have by reference to Fig. 20:

$$dF(ab) = d(a + b) = a + b,\ Fd(ab) = F(ac' + c'b) = a + b = dF(ab);$$

$$dF(abc) = d(ab + bc + ca) = ac' + c'b + ba' + a'c + cb' + b'a,$$

$$Fd(abc) = \text{the sum of the sides of the small triangles in Fig. 20} = dF(abc).$$

This proves (6.4).

If C^s is a chain of K then dC^s is a chain of K_1, referred to as a *subdivided chain.*

(6.5) *If C^s is a cycle of K then dC^s is a cycle of K_1 and conversely.*

Suppose C^s a cycle. Then $FdC^s = dFC^s = 0$ and so dC^s is a cycle. Conversely let dC^s be a cycle. Then $d(FC^s) = 0$ and since $C^t \neq 0$ implies $dC^t \neq 0$, we must have $FC^s = 0$ or C^s is a cycle. This proves (6.5).

(6.6). *Every s-cycle γ_1^s of K_1 is homologous to a subdivided cycle $d\gamma^s$ and if $\gamma^s \sim 0$ in K then $d\gamma^s \sim 0$ in K_1.*

The second part is obvious: if $\gamma^s = FC^{s+1}$, then $d\gamma^s = dFC^{s+1} = F(dC^{s+1}) \sim 0$. Thus we only need to prove $\gamma_1^s \sim d\gamma^s$. Here again it is sufficient to consider elementary subdivisions. Consider first $(6.6)_0$. Referring to Fig. 20, $Fac' = c' + a \sim 0$, hence $c' \sim a$. Therefore in γ_1^0 every vertex of K_1, which is not a vertex of K, may be replaced by a vertex of K. This replaces γ_1^0 by a homologous cycle γ'^0 of K and since $d\gamma'^0 = \gamma'^0$, $(6.6)_0$ follows.

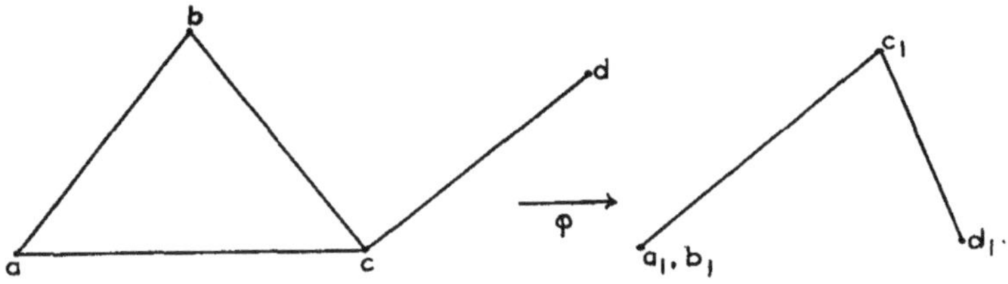

Figure 25

Consider now $(6.6)_1$. If γ_1^1 contains $a'c'$, replace γ_1^1 by $\gamma_1^1 + Fbc'a' \sim \gamma_1^1$. As a consequence $a'c'$ will be eliminated. Repeating the same operation, γ_1^1 will be reduced to a cycle $\gamma'^1 \sim \gamma_1^1$ and consisting solely of simplexes such as bc'. But if $bc' \subset \gamma'^1$ likewise $ac' \subset \gamma'^1$, else $F\gamma'^1$ would contain c', and γ'^1 would not be a cycle. Thus γ'^1 is a sum of chains $(ac' + c'b) = d(ab)$, or $\gamma'^1 = dC^1 \sim \gamma_1^1$. By (6.5) C^1 is a cycle γ^1 and $(6.6)_1$ follows.

Passing now to $(6.6)_2$ if γ^2 is a cycle of K_1 and contains one of the small triangles in Fig. 20 it must contain all the others, for otherwise $F\gamma^2$ would contain a simplex of type $a'b'$. Hence γ^2 is a sum of sums of simplexes such as $d(abc)$, or $\gamma^2 = dC^2$. By (6.5) C^2 is a cycle γ^2. This completes the proof of (6.6).

(6.7) *Application: Every one-cycle of a subdivided tetrahedron boundary bounds.*

Let K be a tetrahedron boundary (see 4.4) and K_1 a subdivision of K. By (4.6) every one-cycle of K is ~ 0. Hence (6.7) follows from (6.6).

7. Simplicial chain-mapping. Let $K = \{\sigma\}$, $K_1 = \{\zeta\}$ be two complexes and let $a,b, \cdots$, and $a_1,b_1, \cdots$, be their vertices. Let φ be an assignment to each vertex of K of a single vertex of K_1 and let the notations be so chosen that $\varphi a = a_1$, $\varphi b = b_1, \cdots$, where repetitions are allowed among the vertices $a_1,b_1, \cdots$. Suppose that φ has the property that if $\sigma = a \cdots b \in K$ then $\zeta = a_1 \cdots b_1 \in K_1$. Under the circumstances let φ be extended to a linear transformation $K \rightarrow K_1$ still written φ, by $\varphi\sigma = \zeta$, where it is understood that if the vertices of ζ are not distinct then $\zeta = 0$ (degenerate simplexes are zero as chains). Thus in Fig. 25: $\varphi ac = a_1c_1$, $\varphi abc = 0$. The linear transformation φ just introduced is said to be *simplicial.*

(7.1) *φ is a chain-mapping.*

Take first $\sigma^1 = ab$. If $a_1 \neq b_1$ then $\varphi Fab = \varphi(a + b) = a_1 + b_1 = F\varphi ab$. If $a_1 = b_1$ then $\varphi Fab = 2a_1 = 0$, $\varphi ab = 0$, $F\varphi ab = 0$. Thus $\varphi F\sigma^1 = F\varphi\sigma^1$ for all σ^1.

Consider now $\sigma^2 = abc$. If a_1,b_1,c_1 are distinct then as above $\varphi Fabc = F\varphi abc$. If say $a_1 = b_1$ then $\varphi abc = 0$, $F\varphi abc = 0$, $\varphi F(abc) = \varphi\ (ab + bc + ca) = 2a_1c_1 = 0$. Thus $\varphi F = F\varphi$ in all cases. Therefore φ is a chain-mapping. It is often called a *projection.*

First example. This example will serve as a preparation to the Jordan curve theorem. We will find there a Jordan curve J in a sphere S, and we will require certain mutual projections of J and its neighborhood on S. It will be necessary for our purpose to implement the system with suitable algebraic machinery.

Some preliminary notions must first be introduced. In any complex, say K already considered, the set of all simplexes having the vertex a or the arc ab as vertex or side, is known as the *star* of a or of ab, written $St\ a$, $St\ ab$. It is convenient to consider also a triangle abc as its own star, written again $St\ abc$. Notice that $St\ a$ includes a, and $St\ ab$ includes ab. The point sets $|St\ a|, \cdots$, will be written $\lambda(a), \cdots$. They are called *star sets,* or merely *stars* whenever no confusion can arise with "star" in the earlier sense. It is an elementary matter to verify that if $a \cdots b \in K$ then

$$\lambda(a) \cap \cdots \cap \lambda(b) = \lambda(a \cdots b).$$

Thus, in Fig. 26, the star of a consists of a, ad, ae, af, ade, aef, afd and $\lambda(a)$ consists of the (open) triangle def.

Let now $\Pi = |K|$, $\Pi_1 = |K_1|$ be two polyhedra, where the complexes are as before. In the applications one of them will contain the other, but the projections which will be needed will have to run "in both directions" from K to K_1 and *vice versa.* To take care of both possibilities we merely assume, therefore, that Π and Π_1 intersect. By the *border of*

Π_1 *in* K, written $B(\Pi_1)$ in K, or merely $B(\Pi_1)$ when the relation to K is clear, is meant the set of simplexes σ of K whose sets $\overline{\lambda(\sigma)}$ meet Π_1. If σ' is a face of σ, then $\lambda(\sigma') \supset \lambda(\sigma)$. Hence whenever $\overline{\lambda(\sigma)}$ meets Π_1, so does $\overline{\lambda(\sigma')}$; that is to say, if σ is in $B(\Pi_1)$, so are all its faces. Thus $B(\Pi_1)$ is a subcomplex of K. It is easily seen that $|B(\Pi_1)|$ is the closure of a neighborhood of the intersection $\Pi \cap \Pi_1$ in Π. The assumption that the two polyhedra intersect is merely introduced to guarantee that the border is nonvacuous. In Fig. 26, $B(\Pi_1)$ in K consists of all the elements of K shown except *ed* and *aed.*

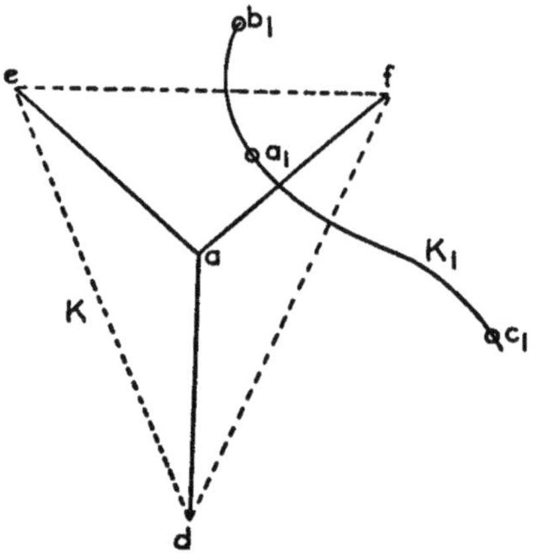

Figure 26

(7.2) We shall say that K *refines* K_1, or is a *refinement* of K_1, under the following circumstances: (a) Π is not contained in Π_1. It is then required that for every set $\overline{\lambda(a)}$ that meets Π_1, i.e. for every vertex a of $B(\Pi_1)$ there is a vertex a_1 of K_1 such that $\overline{\lambda(a)} \cap \Pi_1 \subset \overline{\lambda(a_1)}$. This is manifestly the situation in Fig. 26. (b) Π is contained in Π_1. Then we merely demand that the collection $\{\lambda(a)\}$ refine the collection $\{\lambda(a_1)\}$.

Let K refine K_1 and for each vertex $a,b, \cdots$, of $B(\Pi_1)$ choose a vertex $\varphi a, \varphi b, \cdots$, of K_1 related to it as above. Let the notations be such that $\varphi a = a_1$, $\varphi b = b_1, \cdots$. This time, then, φ is a transformation of the vertices of $B(\Pi_1)$ into those of K_1.

(7.3) φ *defines a simplicial chain-mapping* $B(\Pi_1) \to K_1$.

If $\sigma = a \cdots b \in B(\Pi_1)$, then $\overline{\lambda(a)} \cap \cdots \cap \overline{\lambda(b)} \cap \Pi_1 \subset \overline{\lambda(a_1)} \cap \cdots \cap \overline{\lambda(b_1)} = \overline{\lambda(a_1 \cdots b_1)}$. Hence *St* $a_1 \cdots b_1$ is nonempty, $a_1 \cdots b_1 \in K_1$ and (7.2) follows.

We refer to φ as a *star-projection* $B(\Pi_1) \to K_1$.

In case (b) φ is a chain-mapping $K \to K_1$

Second example. Let $K = \{\sigma\}$ be a subdivision of $K_1 = \{\zeta\}$ and let d be chain-subdivision $K_1 \to K$. Define now $\varphi a = a_1, \cdots$, where $a_1, \cdots$, is a vertex of the simplex ζ of K_1 containing $a, \cdots$ (Fig. 27). If $\sigma = a \cdots b$, then $\bar{\sigma}$ is in a $\bar{\zeta}$. Therefore φ assigns to the vertices of σ vertices of ζ and consequently it can be extended to a simplicial chain-mapping $K \to K_1$. We denote this new chain-mapping by d^{-1} and call it an *inverse* of d. Actually d^{-1} is not unique and the property (7.5) below would only justify the term "a left inverse of d." However for the limited applications to be made the uniqueness of d^{-1} is unimportant and no confusion will arise.

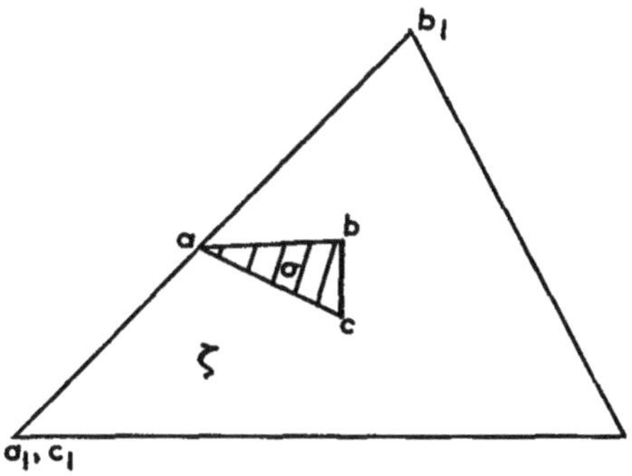

Figure 27

(7.4) *If* $\sigma \in K$ *is in* $\zeta \in K_1$ *and* $d^{-1}\sigma \neq 0$, *then* $d^{-1}\sigma$ *is* ζ *or a face of* ζ.

$$d^{-1} \cdot d = 1. \tag{7.5}$$

The first property is immediate. Regarding the second, let $(7.5)_s$ denote it as applied to the s-chains. Since $d^{-1} \cdot d$ is linear (7.5) need only be verified for the ζ_i^s. Since $d\zeta_i^0 = d^{-1}\zeta_i^0 = \zeta_i^0$, $(7.5)_0$ holds. Assume $(7.5)_s$. By definition $d\zeta_i^{s+1}$ is the sum of the σ_j^{s+1} contained in ζ_i^{s+1}. Each element of the sum is sent by d^{-1} into ζ_i^{s+1} or zero. Hence $d^{-1}d\zeta_i^{s+1} = g\zeta_i^{s+1}$ where $g = 0,1 \bmod 2$. Apply F. Since $(7.5)_s$ is assumed and d, d^{-1} are chain-mappings, we find $Fd^{-1}d\zeta_i^{s+1} = gF\zeta_i^{s+1} = F\zeta_i^{s+1} \neq 0$. Hence $g = 1$ and $(7.5)_{s+1}$, hence (7.5) follows.

(7.6) d^{-1} *is a star-projection* $K \to K_1$ *and conversely every star-projection* $K \to K_1$ *is a suitable* d^{-1}.

Suppose first that K is an elementary subdivision of K_1. A glance at Fig. 20 shows that then in our present notations $\lambda(a) \subset \lambda(a_1)$ if and only if a is in a simplex ζ of $St\, a_1$. By repetition this property holds for any subdivision and this implies (7.6).

(7.7) *Every σ^2 contained in a given ζ^2 is sent by d^{-1} into ζ^2 or zero and the sum of all the σ^2's contained in ζ^2 is sent by d^{-1} into ζ^2.*

The first clause follows from (7.4). By definition $d\zeta^2$ is the sum of the σ^2's which it contains. From (7.5) follows now $d^{-1} \cdot d\zeta^2 = d^{-1}$ (sum of the σ^2's in ζ^2) $= \zeta^2$. This proves (7.7).

(7.8) *If γ_1^s is a cycle of K_1 and $\gamma^s = d\gamma_1^s$ is the subdivided cycle then the relations: $\gamma^s \sim 0$ in K and $\gamma_1^s \sim 0$ in K_1 are equivalent.*

If $\gamma_1^s = FC_1^{s+1}$ then $\gamma^s = d\gamma_1^s = FdC_1^{s+1}$, so the second homology implies the first. If $\gamma^s = FC^{s+1}$ then $\gamma_1^s = d^{-1}d\gamma_1^s = d^{-1}FC^{s+1} = Fd^{-1}C^{s+1}$, so the first homology implies the second. Thus the two are equivalent.

8. Interlocking systems. Let Π,Ω be two polyhedra with $\Omega \subset \Pi$ and let K and L be covering complexes of Π and Ω. Thus $|K| = \Pi$ and $|L| = \Omega$. As an example Π might be a sphere and Ω a Jordan curve contained in the sphere. To link up the properties of chains and cycles with topology one must "interlock" in some manner the complexes K,L and their subdivisions. An adequate mechanism for the purpose will be found in the *interlocking systems*. It belongs to the general method of "simplicial" approximations initiated by L. E. J. Brouwer (around 1910, see IV, V) and is closely related to the method utilized by J. W. Alexander [c] for dealing with the same type of problems.

Let $\{a_i\}$ be the vertices of K. Since a set $\lambda(a_i)$ is an open set of Π and every point of Π is in some $\lambda(a_i)$, $\{\lambda(a_i)\}$ is a finite open covering of Π. Since the simplexes of $St\, a_i$ have in common the point a_i, the mesh $\epsilon(K)$ of the covering does not exceed 2 mesh K. Hence $\epsilon(K) \to 0$ with mesh K. As an open covering of the compactum Π, $\{\lambda(a_i)\}$ has a Lebesgue number which we designate by $\eta(K)$. We recall the basic property of this number (I, 11.3) readily proved in the special case here considered: If $A \subset \Pi$ and diam $A < \eta(K)$, then A is contained in some $\lambda(a_i)$. Similar remarks may be made for any complex K_1, L, $\cdots$ and we will always have the related numbers ϵ, η.

Returning now to the situation described at the beginning we shall take two successive subdivisions K_1,K_2 and L_1,L_2 of K and L and use the following notations:

a_{ni}, b_{ni} ($n = 1,2$) for the vertices of K_n,L_n;

σ_{ni}^s, ζ_{ni}^s for the simplexes of K_n,L_n;

d, d^{-1} for chain-subdivision $K_1 \to K_2$ and an inverse;

d', d'^{-1} for chain-subdivision $L_1 \to L_2$ and an inverse.

Interlocking systems may be utilized for two distinct purposes. When $\Omega = \Pi$, i.e. when K,L cover the same polyhedron Π an interlocking system may serve to correlate the "algebraic" properties of the covering complexes of Π with the topology of the polyhedron Π. On the other hand when Ω is a proper subset of Π an interlocking system may be used to deduce topological relations between Ω and Π from the algebraic properties of the complexes. A noteworthy instance will be found in connection with the Jordan curve theorem. For later purposes it is best to separate the two cases.

First case: $\Omega = \Pi$. Thus here K,L are two covering complexes of the same polyhedron Π. In this case we shall usually not require L_2. Then $\{K_1,L_1,K_2\}$ is an interlocking system whenever L_1 refines K_1 and K_2 refines L_1. The related star-projections are in accordance with the following diagram

$$\begin{array}{c} \xrightarrow{\quad d^{-1} \quad} \\ K_2 \overset{\omega}{\rightarrow} L_1 \overset{\pi}{\rightarrow} K_1 \\ \xleftarrow[\quad d \quad]{} \end{array}$$

Evidently $\pi\omega$ is a star-projection $K_2 \rightarrow K_1$ and hence it may be chosen as d^{-1}. We have then

$$(8.1) \qquad d^{-1} = \pi\omega, \qquad d^{-1}d = 1.$$

In order that the system $\{K_1,L_1,K_2\}$ be interlocking it is manifestly sufficient that

$$\epsilon(L_1) < \eta(K_1),\ \epsilon(K_2) < \eta(L_1).$$

We may choose any subdivision K_1 of K (K_1 may be K itself) and then in succession L_1,K_2 so as to satisfy these two inequalities. Hence there can always be constructed an interlocking system of the type here considered.

Second case: $\Omega \subset \Pi$. Usually this case requires a system of four complexes. We shall then say that $\{K_1,L_1,K_2,L_2\}$ is an interlocking system whenever the following properties are satisfied. Let $B_n = B(\Omega)$ in K_n. Then

Property I. The union of any set $\overline{\lambda(b_{1i})}$ and of all the $\overline{\lambda(a_{2j})}$ which meet it is contained in some set $\lambda(a_{1h})$.

Under the circumstances a_{1h} is a vertex of B_1. Choose for each b_{1i} a definite a_{1h} related to it as above. Since $\overline{\lambda(b_{1i})} \subset \lambda(a_{1h})$, $b_{1i} \rightarrow a_{1h}$ defines a star-projection $\pi_1 : L_1 \rightarrow B_1$.

Property II. L_2 refines K_2 and consequently it refines actually B_2. Denote by π_2 an associated star-projection $L_2 \rightarrow B_2$.

Property III. It is a property of B_2 that if a_{2i} is a vertex of B_2 the intersection $\Omega \cap \overline{\lambda(a_{2i})}$ is nonvoid. It is required that the union of this intersection and of all the sets $\overline{\lambda(b_{2j})}$ that meet it be contained in some set $\lambda(b_{1h})$.

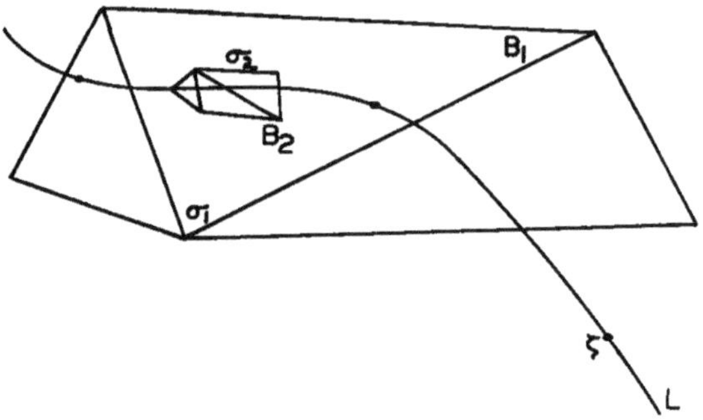

Figure 28

Choose for each a_{2i} of B_2 a definite vertex b_{1h} related to it as above. Since $\overline{\lambda(a_{2i})} \subset \lambda(b_{1h})$, $a_{2i} \to b_{1h}$ defines again a star-projection $\omega : B_2 \to L_1$. All the projections are described in the diagrams

$$\begin{array}{c} \xrightarrow{\qquad\qquad d'^{-1} \qquad\qquad} \\ L_2 \xrightarrow{\ \pi_2\ } B_2 \xrightarrow{\ \omega\ } L_1 \xrightarrow{\ \pi_1\ } B_1 \\ \xleftarrow[\qquad\qquad d' \qquad\qquad]{} \end{array}$$

$$K_2 \xleftarrow{\ d\ } K_1 \qquad \xrightarrow{\ d^{-1}\ }$$

There are some relations however between them already indicated in the first diagram but not yet proved and which we shall now derive.

Suppose first that $\omega a_{2i} = b_{1h}$, $\pi_1 b_{1h} = a_{1k}$. Then by Property I: $\lambda(a_{1k}) \supset \overline{\lambda(b_{1h})}$ and so $\lambda(a_{1k})$ meets Ω or a_{1k} is a vertex of B_1. Since $\omega a_{2i} = b_{1h}$, $\lambda(a_{2i})$ meets $\lambda(b_{1h})$ and hence, by Property I, $\overline{\lambda(a_{2i})}$ is contained in $\lambda(a_{1k})$. Thus $a_{2i} \to a_{1k}$ is a suitable determination of d^{-1} on the vertices of B_2. On the other vertices of K_2 we determine it in arbitrary manner. Thus on B_2 we shall have $d^{-1} = \pi_1 \omega$.

Suppose now that $\pi_2 b_{2h} = a_{2i}$ and $\omega a_{2i} = b_{1k}$. Thus by Property II: $\overline{\lambda(b_{2h})} \subset \lambda(a_{2i})$ and by Property III: $\Omega \cap \overline{\lambda(a_{2i})}$ together with $\overline{\lambda(b_{2h})}$ is contained in $\lambda(b_{1k})$. Hence $b_{2h} \to b_{1k}$ is a suitable determination of d'^{-1} and we have $\omega \pi_2 = d'^{-1}$.

To sum up we have here

(8.2) $$d^{-1} = \pi_1\omega \text{ on } B_2,\ d^{-1}d = 1,\ d'^{-1} = \omega\pi_2,\ d'^{-1}d' = 1.$$

Since K_2 is a subdivision of K_1 any σ_{2i} is in some σ_{1j} and $\overline{\lambda(\sigma_{2i})} \subset \overline{\lambda(\sigma_{1j})}$. Hence if $\overline{\lambda(\sigma_{2i})}$ meets Ω so does $\overline{\lambda(\sigma_{1j})}$. Therefore if $\sigma_{2i} \in B_2$ then $\sigma_{1j} \in B_1$. It follows that $|B_2| \subset |B_1|$. Since $\Omega \subset \Pi$ every point of Ω is in some σ_2 and hence in $\overline{\lambda(\sigma_2)}$ and this $\sigma_2 \in B_2$. Thus

(8.3) $$\Omega \subset |B_2| \subset |B_1|.$$

We also note that the points of $|B_2|$ are not further than mesh K_2 from Ω. Hence K_2 may be chosen such that all the points of $|B_2|$ are nearer to Ω than any preassigned $\alpha > 0$.

In the case $\Omega \subset \Pi$ the interlocking system may again be constructed beginning with any subdivision K_1 of K (K_1 may be K itself) and choosing in succession L_1, K_2, L_2 so as to satisfy the three inequalities

$$\epsilon(L_1) < \tfrac{1}{3}\,\eta(K_1),$$

$$\epsilon(K_2) < \tfrac{1}{3} \inf \{\eta(K_1),\ \eta(L_1)\},$$

$$\epsilon(L_2) < \tfrac{1}{3} \inf \{\eta(L_1),\ \eta(K_2)\}.$$

The construction is manifestly always possible in the asserted manner.

§3. THE JORDAN CURVE THEOREM.

9. We recall that a Jordan curve is any homeomorph of the circumference. The central theorem regarding such curves is:

(9.1) **Theorem.** *A Jordan curve J in the plane Π decomposes $\Pi - J$ into two regions whose common boundary is J.*

Of the two regions into which J decomposes its complement one U is bounded and the other V is not. The first is the *interior* of J, the second the *exterior* of J. Notice that $\bar{U} = U \cup J$ is compact while $\bar{V}$ is not.

It is of much interest to investigate the relation of J to 2-cells and 2-spheres. Since a 2-cell E^2 is topologically equivalent to Π, a Jordan curve $J \subset E^2$ divides $E^2 - J$ into two components U,V with the common boundary J. Of these two components one U has a compact closure $\bar{U} \subset E^2$ and U is again called the interior of J and V its exterior.

The behavior as to a 2-sphere S is somewhat different. First of all if $J \subset S$ it must be a proper subset. For J, like the circumference, is disconnected by the removal of any two points, and S is not; hence $J \neq S$. Hence $S - J$ contains at least one point P. Choosing now as S the Euclidean sphere in 3-space $\mathfrak{E}^3$, we see that the stereographic projection of $S - P$ from P onto the plane Π tangent at the antipode of P is topological. From this follows readily that (9.1) is equivalent to

(9.2) **Theorem.** *A Jordan curve J in the 2-sphere S divides $S - J$ into two regions whose common boundary is J.*

This last theorem is the one which we propose to prove. An important complement is the

(9.3) **Theorem of Schoenflies.** *If U is either one of the components of $S - J$ then $\bar{U} = U \cup J$ is a closed 2-cell.*

The theorem of Jordan and the theorem of Schoenflies are often stated together as the

(9.4) **Theorem of Jordan-Schoenflies.** *Let the Jordan curve lie in the Euclidean plane Π. Then $\Pi - J$ consists of two components, one of them U bounded, the other V unbounded. The set $U \cup J$ is a closed 2-cell, and J is the common boundary of U and V.*

The Schoenflies theorem is also a corollary of the classical

(9.5) **Conformal Mapping Theorem.** *Under the same conditions as in the Jordan-Schoenflies theorem, there is a topological mapping f of $\bar{U}$ onto a closed circular region which is conformal at all points of U and also at all points of J which are contained in some analytical subarc of J.*

The initial proof of his theorem given by Schoenflies is strictly topological and straightforward enough. The proof of the conformal mapping theorem, while of necessity analytical, is comparatively simple and is found in texts on complex variables. [See Caratheodory's Cambridge Tract: Conformal representation (1932).] The reader is therefore referred to these books for the proof of (9.5). The Schoenflies theorem will be accepted here as a consequence of (9.1) and (9.5) combined.

Evidently also:

(9.6) *The Jordan curve theorems* (9.1), (9.2), (9.3) *have topological character.*

10. Remark on connectedness. In the treatment of (9.2), and more generally in questions on subsets of spheres it is convenient to reduce whenever possible general connectedness to *arc-wise connectedness.* In this connection we dispose of

(10.1) *A connected open subset (region) of the sphere S is arc-wise connected.*

Let U be the region and set $G = S - U$. Take a point $P \in U$ and let H be the set of points of U which may be joined to P (and hence to one another) by a broken line, (a closed arc made up of a finite number of closed circular arcs). The set H is arc-wise connected. If $Q \in \bar{H} \cap U$, some circular region $\mathfrak{S}(Q,\epsilon) \subset U$ contains a point $R \in H$. Therefore $\mathfrak{S}(Q,\epsilon) \subset H$, and hence $Q \in H$. Thus H is closed in U. If $Q \in H$ some

$\mathfrak{S}(Q,\epsilon) \subset H$ also, hence H is open in U. Since H is connected and both open and closed in the connected set $U,H = U$ and (10.1) follows.

11. Algebraic preparation. Since in the proof of the Jordan curve theorem one may replace the sphere S by any homeomorph one may assume that S is the boundary of a tetrahedron. Let K be the complex made up of the triangles, edges and vertices of the boundary of the tetrahedron. The Jordan curve J is decomposed into three arcs by three subdivision points and these arcs and points make up a complex L. Since $J \subset S$ we may form an interlocking system $\{K_1,L_1,K_2,L_2\}$ as in (8) second case. The general notations of (8) will be used here. In par-

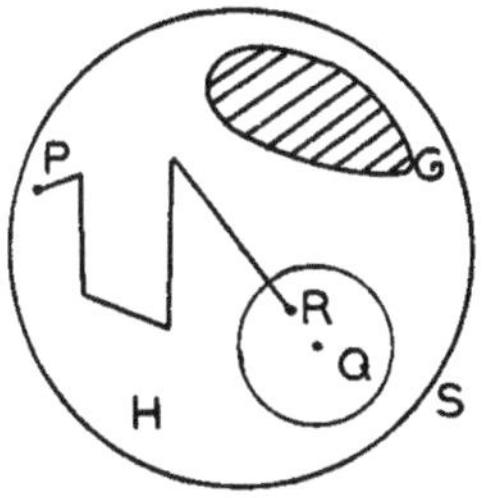

Figure 29

ticular B_n will denote the border of J in K_n. In addition C_n, D_n will denote chains of K_n. No other chains than γ_n^1 will occur in L_n. Recall also that $n = 1,2$.

(11.1) *Every σ_{ni}^1 is incident with two triangles of K_n.*

This property is verified directly for K itself. Referring to Fig. 20 it is seen that if it holds for any complex it holds for its elementary subdivision and hence for any subdivision. Since it holds for K it holds for its subdivision K_n.

(11.2) *Every one-cycle of K_n bounds.* (6.7).

Set now

(11.3) $\gamma_n^1 = \Sigma\zeta_{ni}^1, \; \gamma_n^2 = \Sigma\sigma_{ni}^2.$

(11.4) *The only one-cycle of L_n which is $\neq 0$ is γ_n^1.*

For L_n is isomorphic with the complex resulting from a subdivided circumference and so (11.4) follows from (4.3) which is manifestly unaffected by an isomorphism.

(11.5) *The only 2-cycle of K_n which is $\neq 0$ is γ_n^2.*

Let d_n be the operation of chain-subdivision $K \rightarrow K_n$ and let γ_o^2 be the sum of the triangles of K. Thus $\gamma_n^2 = d_n\gamma_o^2$. By (4.5) γ_o^2 is the only 2-cycle $\neq 0$ in K. Hence by (6.6) a 2-cycle $\gamma_n'^2$ of K_n which is $\neq 0$ is

$\sim \gamma_n^2$. Thus $\gamma_n'^2 - \gamma_n^2$ is a boundary. Since K_n contains no 3-chains, $\gamma_n'^2 - \gamma_n^2 = 0$ which is (11.5).

(11.6) *Let C_n^2 be a chain of K_n and U any component of $S - |FC_n^2|$. If C_n^2 contains one of the triangles of K_n which meet U then it contains them all.*

Let $\{\sigma_{ni}'^2\}$ be the triangles in question and suppose $\sigma_{ni}'^2 \subset C_n^2$, $\sigma_{nj}'^2$ not in C_n^2. Take points x,y of U respectively in $\sigma_{ni}'^2, \sigma_{nj}'^2$. We may draw in U an arc μ from x to y crossing no vertices. As μ is described from x to y there will be reached a last point z on the closure $\bar{\sigma}_{nk}'^2$ of a $\sigma_{nk}'^2 \subset C_n^2$ (Fig. 30). The point z will be on a σ_n^1 of K_n incident with $\sigma_{nk}'^2 \subset C_n^2$ and with $\sigma_{nl}'^2$ not in C_n^2. Hence σ_n^1 is in FC_n^2, a contradiction since $z \in S - |FC_n^2|$. This proves (11.6).

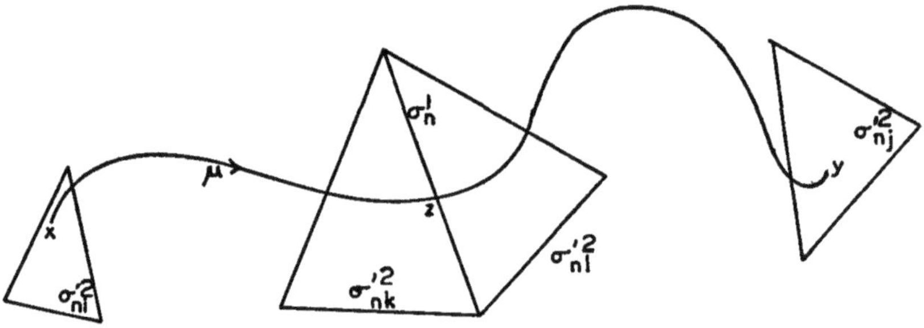

Figure 30

Remark. If C_n^2 is a cycle $U = S$ and (11.6) reduces to (11.5).

From the definition of the cycles we find at once

$$\gamma_2^1 = d'\gamma_1^1; \qquad \gamma_2^2 = d\gamma_1^2. \tag{11.7}$$

Since L_n contains no two-chain and $\gamma_n^1 \neq 0$ we also have

$$\gamma_n^1 \not\sim 0. \tag{11.8}$$

Set now $\delta_n^1 = \pi_n \gamma_n^1$ so that δ_n^1 is a cycle of the border B_n of J in K_n. Then

$$\delta_1^1 = d^{-1}\delta_2^1. \tag{11.9}$$

We have in fact

$$\begin{aligned} d^{-1}\delta_2^1 &= d^{-1}\pi_2\gamma_2^1 = \pi_1\omega\pi_2\gamma_2^1 \\ &= \pi_1 d'^{-1}\gamma_2^1 = \pi_1 d'^{-1}d'\gamma_1^1 = \pi_1\gamma_1^1 = \delta_1^1. \end{aligned}$$

(11.10) *The interlocking system may and shall be so chosen that $\delta_n^1 \not\sim 0$ in B_n.*

Without any special choice we have $\delta_2^1 \not\sim 0$. For $\delta_2^1 \sim 0$ implies $\omega\delta_2^1 =$

$\gamma_1^1 \sim 0$ which is untrue. Since we may replace K_1 by K_2 as the initial complex we may choose the interlocking system so that $\delta_1^1 \nsim 0$ also.

(11.11) *There are exactly two chains* C_n^2, D_n^2 *such that*

$$\delta_n^1 = FC_n^2 = FD_n^2, \qquad C_n^2 + D_n^2 = \gamma_n^2,$$

and these two chains have no common triangles.

Since δ_n^1 is in K_n it bounds in K_n. Hence $\delta_n^1 = FC_n^2$. If we set $D_n^2 = C_n^2 + \gamma_n^2$, then since γ_n^2 is a cycle $\delta_n^1 = FD_n^2$. Since the sum of C_n^2, D_n^2 is γ_n^2 which is the sum of all the triangles of K_n, no σ_{nt}^2 may be in both C_n^2 and D_n^2, for in that case it would not be in γ_n^2.

Suppose now $FC_n'^2 = \delta_n^1$. Since $\delta_n^1 \neq 0$, $C_n'^2$ is neither zero nor γ_n^2 and hence there is a σ_{nt}^2 in $C_n'^2$. It must also be in one of C_n^2, D_n^2, say in the first. Then $C_n^2 + C_n'^2$ is a two-cycle which does not contain σ_{nt}^2 and so is not γ_n^2. Hence $C_n^2 + C_n'^2 = 0$, $C_n^2 = C_n'^2$. This proves (11.11).

(11.12) *The notations may be so chosen that*

$$d^{-1}C_2^2 = C_1^2, \qquad d^{-1}D_2^2 = D_1^2.$$

We have in fact

$$\delta_1^1 = d^{-1}\delta_2^1 = d^{-1}F(C_2^2) = d^{-1}F(D_2^2) = F(d^{-1}C_2^2) = F(d^{-1}D_2^2),$$

$$(d^{-1}C_2^2) + (d^{-1}D_2^2) = (d^{-1}\gamma_2^2) = \gamma_1^2.$$

This proves that $d^{-1}C_2^2$ and $d^{-1}D_2^2$ are the same as C_1^2 and D_1^2 in some order. It follows that the notations may be so chosen that (11.12) holds.

§4. Proof of the Jordan Curve Theorem.

12. All the preliminary algebraic machinery having been developed, the proof proper will proceed quite swiftly.

(12.1) $S - J$ *has at least two components.*

Referring to (11.10) neither C_n^2 nor D_n^2 of (11.11) can be in B_n. Hence each contains triangles not in B_n. Choose a point x in a σ_{1t}^2 of C_1^2 not in B_1 and a point y in a σ_{1j}^2 of D_1^2 likewise not in B_1. To prove (12.1) we merely need to show that

(12.2) x *and* y *are in distinct components of* $S - J$.

If (12.2) does not hold there may be drawn an arc ξ from x to y in $S - J$, and ξ will be at a positive distance α from J. It follows that if we take K_2 of mesh $< \alpha$, B_2 will not meet ξ and so ξ will not meet $|\delta_2^1|$. Then x and y will be in the same component U of $S - |\delta_2^1.|$ We shall show that this leads to a contradiction.

Suppose in fact $x \in \sigma_{1t}^2$, $x \in \bar{\sigma}_{2j}^2$. This can only be if $\sigma_{2j}^2 \subset \sigma_{1t}^2$. Now d^{-1} sends every $\sigma_{2h}^2 \subset \sigma_{1t}^2$ into σ_{1t}^2 or into zero and it sends their sum into σ_{1t}^2. This implies that if σ_{2h}^2 is not in σ_{1t}^2 then its image under d^{-1} is not

σ^2_{1i}. It follows then from $d^{-1}C^2_2 = C^2_1$ and since σ^2_{1i} is in C^2_1, that some σ^2_{2h} in σ^2_{1i} is imaged by d^{-1} into σ^2_{1i}, and σ^2_{2h} is therefore in C^2_2. Since δ^1_2 is in B_2 and $|B_2| \subset |B_1|$ while σ^2_{1i} is not in B_1, σ^2_{1i} does not meet $|\delta^1_2|$. Hence in view of (11.6) and since σ^2_{2h} is in C^2_2, every σ^2_{2k} contained in σ^2_{1i} is in C^2_2. Similarly if $y \in \sigma^2_{1j}$ then σ^2_{1j} does not meet $|\delta^1_2|$ and every σ^2_{2l} in σ^2_{1j} is in D^2_2.

We conclude then that the component U of $S - |\delta^1_2|$ which, by hypothesis, contains both x and y, contains also simplexes $\sigma^2_{2k} \subset \sigma^2_{1i}$ and $\sigma^2_{2l} \subset \sigma^2_{1j}$. Hence σ^2_{2k} and σ^2_{2l} must be both in C^2_2 or both in D^2_2. Since

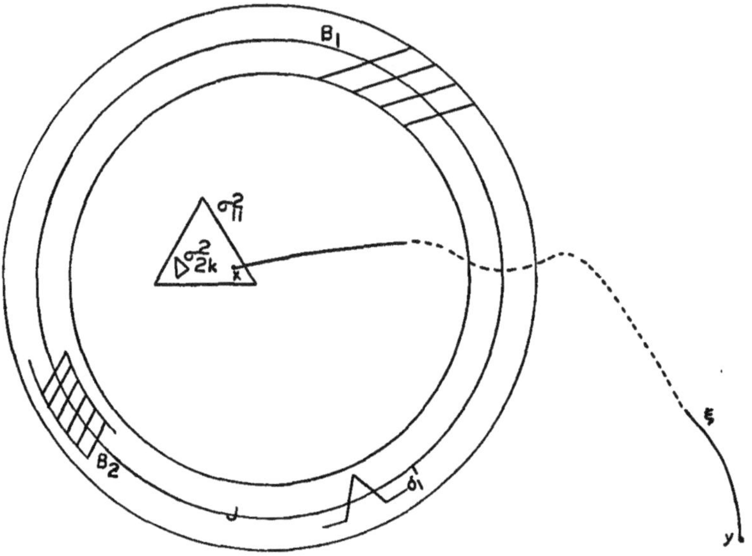

Figure 31

C^2_2 and D^2_2 have no common simplexes and $\sigma^2_{2k} \subset C^2_2$, $\sigma^2_{2l} \subset D^2_2$, we have arrived at a contradiction. This proves (12.2) and hence also (12.1).

13. (13.1) *$S - J$ has at most two components.*

Let U be a component of $S - J$. Since $|B_1| \to J$ when mesh $K_1 \to 0$, we may choose K_1 such that U is not in $|B_1|$ and since the latter contains $|B_2|$, U will then likewise not be in $|B_2|$. Since the σ^2_{2i} are connected those which meet U and are not in B_2, are in U. Let C^2_2 be their sum.

(13.2) *$\bar{\gamma}^1_2 = FC^2_2$ is in B_2.*

Any σ^1_2 in U but not in B_2 is the common side of two triangles σ^2_{2i}, σ^2_{2j} neither of which is in B_2 since otherwise B_2 would contain σ^1_2. The closures of the two triangles are connected and meet U. Hence σ^2_{2i}, σ^2_{2j} are in U and therefore in C^2_2. As a consequence σ^1_2 is not in FC^2_2 and this proves (13.2).

Let $C_1^2 = d^{-1}C_2^2$. By assumption there are points of U not in $|B_1|$ and hence there is a σ_{1i}^2 which meets U, consequently is in U, but not in B_1. The σ_{2j}^2 contained in σ_{1i}^2 are in C_2^2 and their sum and only their sum is imaged by d^{-1} into σ_{1i}^2. Hence the latter is in C_1^2 and so $C_1^2 \neq 0$. Since d^{-1} is a chain-mapping $\bar{\gamma}_1^1 = d^{-1}\bar{\gamma}_2^1 = FC_1^2$.

On the other hand $\bar{\gamma}_1^1 = \pi_1\omega\bar{\gamma}_2^1$. Since $\bar{\gamma}_2^1$ is in B_2 its image $\omega\bar{\gamma}_2^1$ is in L_1 and so $\omega\bar{\gamma}_2^1 = \alpha\gamma_1^1$, $\alpha = 0$ or 1 (11.4). Hence $\bar{\gamma}_1^1 = \alpha\pi_1\gamma_1^1 = \alpha\delta_1^1$.

To sum up:

(13.3) *The chain* C_1^2 *is not zero. Its triangles are all in* U *and its boundary is* $\bar{\gamma}_1^1 = \alpha\delta_1^1$.

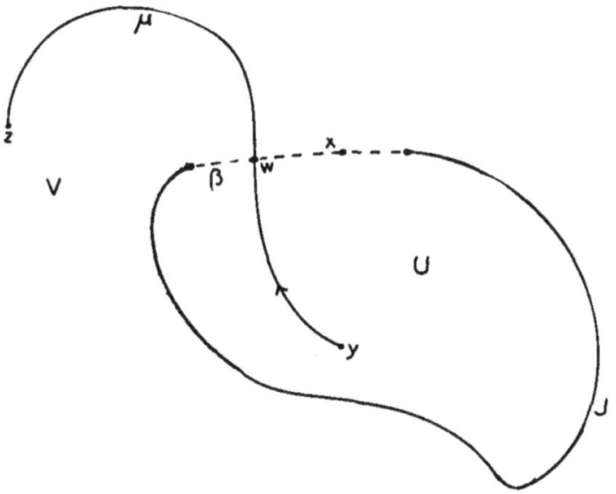

Figure 32

Suppose now that $S - J$ has three distinct components U,V,W. Choose K_1 such that none of the three is in $|B_1|$. Form C_1^2, $\bar{\gamma}_1^1$ as above for U and analogues $C_1'^2$, $\bar{\gamma}_1'^1$ for V with $\bar{\gamma}_1'^1 = \alpha'\delta_1^1$. Then $\bar{\gamma}_1^2 = \epsilon C_1^2 + \epsilon' C_1'^2$ will have no triangle in W, and its boundary is $(\epsilon\alpha + \epsilon'\alpha')\,\delta_1^1$. We may choose ϵ,ϵ' not both 0 mod 2 such that $\epsilon\alpha + \epsilon'\alpha' = 0$ mod 2. Then $\bar{\gamma}_1^2$ will be a 2-cycle $\neq 0$ and yet not containing the triangles in W. Thus $\bar{\gamma}_1^2 \neq 0$, $\bar{\gamma}_1^2 \neq \gamma^2$. Since this contradicts (11.5) it proves (13.1).

From (12.1) and (13.1) follows:

(13.4) $S - J$ *has exactly two components.*

We shall now prove:

(13.5) *If* H *is a closed arc in* S *then* $S - H$ *is connected.*

Take now for L the arc and its two end-points. Since L has no one-cycle the reasoning leading to (13.4) yields here that C_1^2 is a 2-cycle whose triangles are all in a component U of $S - H$. Hence there can be no component of $S - H$ other than U, since then as above $C_1^2 \not\sim \gamma_1^2$, 0.

(13.6) *The two components U,V of $S - J$ have J as their common boundary.*

Let $x \in J$ and corresponding to any $\epsilon > 0$ let β be an open arc of J containing x and of diameter $< \epsilon$. Since $H = J - \beta$ is a closed arc, $S - H$ is connected. Hence if $y \in U$, $z \in V$ there is an arc μ joining the two points in $S - H$. Since U,V are distinct components of $S - J$, μ must cross J and hence it must cross β since $\mu \subset S - H$. As μ is followed from y to z its first crossing w of β is in the boundary $\mathfrak{B}U$ of U. Hence $\mathfrak{B}U$ is dense in J and since it is closed in J, $\mathfrak{B}U = J$. Similarly $\mathfrak{B}V = J$.

Propositions (13.4) and (13.6) together prove the Jordan curve theorem.

5. SOME ADDITIONAL PROPERTIES OF COMPLEXES.

14. We shall only consider the few complements necessary for a suitable treatment of graphs and surfaces.

Let $K = \{\sigma\}$ be a complex of the type of the present chapter. Suppose that there may be found R^s cycles $\{\gamma_i^s\}$, $i = 1,2, \cdots, R^s$ of K such that:

(a) every s-cycle γ^s of K satisfies a relation

$$\gamma^s \sim \Sigma g_i\gamma_i^s, \text{ mod } 2;$$

(b) the γ_i^s are independent (understood as to bounding), i.e. they do not satisfy any homology

$$\Sigma g_i\gamma_i^s \sim 0, \text{ mod } 2,$$

unless every $g_i = 0$ mod 2.

Under the circumstances the number R^s is called the *sth Betti number* of K. (Later, in the following chapters, when there will be others the same number will be called a Betti number mod 2 and written R_2^s.) We shall also write $R^s(K)$ for R^s.

Examples. It is implicit in our previous results that certain Betti numbers have the values indicated below:

Subdivided Jordan curve: $R^1 = 1$.

Subdivided arc: $R^1 = 0$.

Subdivided tetrahedron boundary: $R^1 = 0$, $R^2 = 1$.

We shall now prove the following special case of an important theorem due to J. W. Alexander. The general case is treated in (V,8).

(14.1) **Theorem.** *The Betti numbers of K are topologically invariant* *in the sense that if the two polyhedra $|K|$, $|L|$ are homeomorphic then the complexes K,L have the same Betti numbers.*

This implies for instance that no matter how a sphere S is triangulated the Betti numbers of the resulting complex are the same as for the tetrahedron boundary: $R^1 = 0$, $R^2 = 1$.

If $|L|$ is mapped onto $|K|$ by the homeomorphism $|L| \to |K|$, L goes into an isomorph L_1 with the same Betti numbers. Thus we may suppose $|L| = |K|$, or that K,L are merely different triangulations of the same polyhedron. This being the case we construct an interlocking system $\{K_1,L_1,K_2\}$ as in (8) first case. The situation is precisely the one described there and the same notations will be used here.

Combining the linearity of chain-subdivision with $(6.6)_s$ and (7.8) we find $\rho = R^s(K_n) = R^s(K)$, $\rho' = R^s(L_n) = R^s(L)$. Let $\gamma_{11}^s, \cdots, \gamma_{1\rho}^s$ be independent cycles of K_1 and set $\gamma_{2t}^s = d\gamma_{1t}^s$. If $\rho' < \rho$ the cycles $\{\omega\gamma_{2t}^s\}$ are not independent. Since a chain-mapping preserves the homologies between cycles we find that the cycles $\pi\omega\gamma_{2t}^s$ which are merely the $d^{-1}d\gamma_{1t}^s = \gamma_{1t}^s$ are likewise not independent. This contradiction shows that $\rho' \geqq \rho$. By symmetry $\rho \geqq \rho'$, hence $\rho = \rho'$ or $R^s(K) = R^s(L)$. This proves the theorem.

Since R^s depends solely upon the polyhedron $\Pi = |K|$, we may define it as the *sth Betti number* of Π and write it accordingly $R^s(\Pi)$.

15. A few properties of the Betti numbers will now be stated. The proofs are very simple, and with the exception of one, they are omitted.

*(15.1) *Upon removing the triangles of K there is left a subcomplex L which is a graph, and $R^0(K) = R^0(L)$.*

*(15.2) *If K is connected any two vertices are homologous and hence $R^0(K) = 1$.*

*(15.3) *Let α^s be the number of s-simplexes of K. The number $\chi(K) = \alpha^0 - \alpha^1 + \alpha^2$, called the characteristic of K, is unchanged under subdivision.*

An important property known as the *Euler-Poincaré relation* asserts that

$$\chi(K) = R^0 - R^1 + R^2. \tag{15.4}$$

This relation implies then, in view of (14.1) that

(15.5) *$\chi(K)$ is a topological invariant of the set $|K|$.*

The Euler-Poincaré relation will be dealt with fully in the next chapter. At the present time we only need it for connected graphs and surfaces and we shall prove it separately for these two cases.

A property of the zero-chains. Let us associate with any zero-chain

$C^0 = \Sigma g_i \sigma_i^0$ the number $\kappa(C^0) = \Sigma g_i$ (mod 2). (In the terminology of the next chapter this is a *Kronecker index* mod 2.) We shall need

(15.6) *If K is connected a n.a.s.c. for* $C^0 \sim 0$ *is* $\kappa(C^0) = 0$.

If $C^0 = F\Sigma g_i \sigma_i^1$, then

$$\kappa(C^0) = \Sigma g_i \kappa(F\sigma_i^1) = 0,$$

for if $\sigma^1 = ab$, $\kappa(F\sigma^1) = \kappa(a + b) = 0$.

Conversely suppose $\kappa(C^0) = 0$. If $C^0 = \Sigma h_j \sigma_j^0$ then $\sigma_j^0 \sim \sigma_1^0$ (15.2). Hence

$$C^0 \sim \sigma_1^0 \Sigma h_j = \kappa(C^0)\sigma_1^0 = 0.$$

Betti numbers of the closed two-cell. They are those of a closed triangle and readily found to be $R^0 = 1$, $R^1 = R^2 = 0$. Hence every one-cycle of a closed two-cell bounds.

16. Graphs. Let us observe that since a graph contains no two-chains, ~ 0 and $= 0$ are the same thing for its one-cycles. Thus $R^1(K)$ is merely the maximum number of linearly independent one-cycles, where linear dependence is understood in the ordinary sense for linear forms with coefficients integers mod 2.

Suppose that γ^1 is a one-cycle and let a be a vertex of γ^1, ab a simplex of γ^1 with the vertex a, etc.; there arises thus a sequence $ab, bc, cd, \cdots$ of simplexes of γ^1. In this manner we may never reach a stopping point f, else f would be in $F\gamma^1$ which is zero. Thus in our sequence we must recross some vertex. Let a be such that some sequence $ab, bc, \cdots, fa$ exists made up of simplexes of γ^1 with vertices $a, \cdots, f$ all distinct. Then $\delta^1 = ab + \cdots + fa$ is a cycle. Such a cycle will be referred to as an *elementary* cycle. (It is generally named "circuit" but this term is used in another sense later.) Now $\gamma'^1 = \gamma^1 + \delta^1$ is a cycle in which $ab, \cdots, fa$ are not found any more. By repeating the process there is obtained a finite set of elementary cycles $\delta_1^1, \cdots, \delta_r^1$ such that no two have elements in common and that γ^1 is their sum. As a consequence we may state:

(16.1) *If* γ^1 *contains* σ^1 *then* $|K - \sigma^1|$ *is still connected.*

It is sufficient to show that any two vertices a,b may be joined by an arc in $|K - \sigma^1|$, i.e. that a,b are in a connected subgraph M of $K - \sigma^1$. The vertices a,b may be joined in K by an arc which is a set $|L|$ on which there is a subgraph L of K. If L does not contain σ^1 then we may choose $M = L$. Let L contain σ^1 and let σ^1 be, say, in δ^1. If N is the subcomplex made up of the simplexes of δ^1 other than σ^1 together with their vertices then $M = L \cup N - \sigma^1$ manifestly answers the question.

Consider now a maximal set of independent elementary cycles

$\delta_1^1, \cdots, \delta_\rho^1$. Clearly $\rho \leq R^1$. On the other hand since every one-cycle is a linear combination of the δ_i^1, $\rho \geq R^1$. Thus $\rho = R^1$. This yields the following noteworthy geometric interpertation of R^1, geometric since elementary cycles have an obvious geometric meaning as cycles of a simple polygon:

(16.2) *The number R^1 of the graph K is equal to the maximum number of its linearly independent elementary cycles.*

Example. Let K consist of the edges and vertices of a tetrahedron $abcd$. Then the three elementary cycles $\delta_1^1 = Fdab$, $\delta_2^1 = Fdbc$, $\delta_3^1 = Fdca$, are independent since each contains an edge not found in the other two. If $\delta_4^1 = Fabc$, then $\delta_4^1 = \delta_1^1 + \delta_2^1 + \delta_3^1$. Since in the tetrahedron boundary every $\gamma^1 \sim 0$, γ^1 is a sum of cycles $\delta_1^1, \delta_2^1, \delta_3^1$. Therefore $R^1 = 3$. We find here: $\chi(K) = -2 = 1 - R^1(K)$ so that the Euler-Poincaré relation holds.

The Euler-Poincaré relation for a connected graph. It reads actually

$$\chi(K) = 1 - R^1(K) \tag{16.3}$$

and our next task is to prove it.

A connected graph without one-cycles is known as a *tree*. When K is a tree, $R^1(K) = 0$, and so the Euler-Poincaré relation for a tree reads

$$\chi(K) = 1. \tag{16.4}$$

We first prove this special relation. When K is a tree upon starting from any vertex a, then following successive arcs one will never return to a, since otherwise K would contain some one-cycle. Thus one will come to a stopping point b which can only be the end-point of one and only one arc σ^1 of K. If b and σ^1 are both suppressed, the situation is unchanged and $\chi(K)$ remains the same. This process ends when K is reduced to a single vertex. At that stage $\chi(K) = 1$, hence (16.4) holds for the initial tree.

Take now any connected graph and let $R^1(K) = n$. Designate (16.3) for n by $(16.3)_n$. Thus we have proved $(16.3)_0$. We assume now $(16.3)_{n-1}$ and prove $(16.3)_n$. Let $B = \{\delta_1^1, \cdots, \delta_n^1\}$ be a base for the one-cycles of K and let σ^1 be a simplex of δ_n^1. Denote also by ϵ_i the coefficient of σ^1 in δ_i^1. Setting $\delta_i'^1 = \delta_i^1 + \epsilon_i\delta_n^1$, so that $\delta_i'^1$ does not contain σ^1, we have in $B' = \{\delta_1'^1, \cdots, \delta_{n-1}'^1, \delta_n^1\}$ a new base for the one-cycles. For $\delta_i^1 = \delta_i'^1 + \epsilon_i\delta_n^1$, hence all the one-cycles of K may be expressed in terms of B'. On the other hand, assume a relation

$$\sum_{i<n}\lambda_i\delta_i'^1 + \lambda_n\delta_n^1 = 0. \tag{16.5}$$

Since only δ_n^1 contains σ^1, we must have $\lambda_n = 0$. Then (16.5) yields

$$\sum_{i<n} \lambda_i \delta_i^1 + \mu \delta_n^1 = 0.$$

Since B is a base every $\lambda_i = 0$ for $i < n$ and then B' is a base.

Let now $K - \sigma^1 = L$ so that L is connected, and let γ^1 be a one-cycle of L. Since γ^1 is also a cycle of K, there is a relation

$$\gamma^1 \equiv \lambda_1 \delta_1'^1 + \cdots + \lambda_{n-1} \delta_{n-1}'^1 + \mu \delta_n^1. \tag{16.6}$$

Among the cycles γ^1, $\delta_i'^1$, δ_n^1 only the last contains σ^1. Hence (16.6) implies $\mu = 0$, and so γ^1 is a linear combination of the $\delta_i'^1$. Since the latter are linearly independent, they form a base for the one-cycles of L. Hence $R^1(L) = n - 1 = R^1(K) - 1$. Also $\chi(L) = \chi(K) + 1$. Under the hypothesis that $(16.3)_{n-1}$ holds, we have $\chi(L) = 1 - R^1(L)$ or $\chi(K) + 1 = 1 - (R^1(K) - 1)$, which proves $(16.3)_n$ and hence (16.3).

§6. CLOSED SURFACES. GENERALITIES.

17. The natural definition of a *closed surface* Φ as it arises in the applications is as follows: Φ is a connected compactum such that each point has for neighborhood a two-cell. Owing to compactness Φ may be covered with a finite number of two-cells.

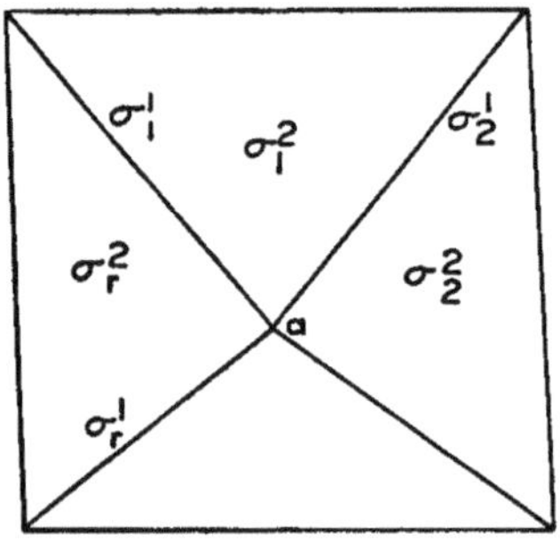

Figure 33

Starting, on the other hand, from a complex $K = \{\sigma\}$ one may impose certain natural conditions of smoothness making $|K|$ a surface. The following two conditions are sufficient.

(17.1) *Every arc σ^1 of K is incident with exactly two triangles.*

It is a consequence of (17.1) that if a is any vertex of K then the triangles and edges having a as vertex fall into a finite number of circular systems $\sigma_1^1 \sigma_1^2 \sigma_2^1 \cdots \sigma_r^2 \sigma_1^1$ (Fig. 33) such that in each system consecutive simplexes are incident. The second condition is:

(17.2) *About each vertex of K there is exactly one circular system of the type just described.*

It is clear that each point has now for neighborhood a two-cell. Thus $|K|$ is a closed surface.

A complex which satisfies conditions (17.1) and (17.2) is known as a *combinatorial absolute two-manifold.*

*(17.3) **Theorem.** *Every closed surface may be triangulated and the resulting complex is a combinatorial absolute two-manifold.*

The proof, due to T. Rado, will be found in his paper of *Szeged Acta* (1925), pp. 101-121.

§7. Closed Surfaces. Reduction to a Normal Form.

18. We propose now to show that each closed surface Φ may be reduced to a certain type which is unique for the class of all homeomorphs of Φ. The surface Φ is taken triangulated as a definite complex $K = \{\sigma\}$. It is with K that we operate at the outset.

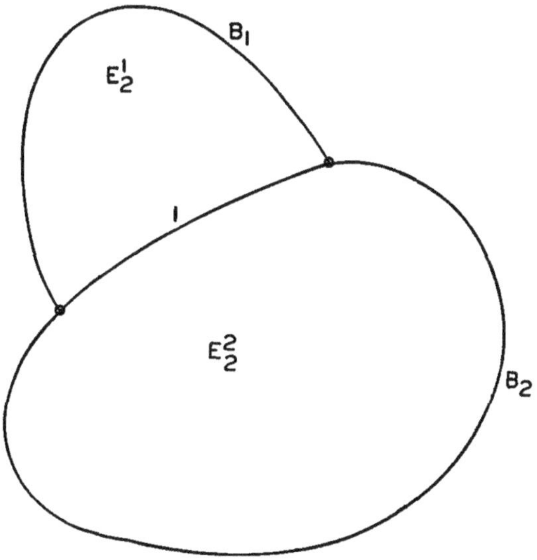

Figure 34

Since Φ is connected the proof of (11.5) is applicable and so we have:

(18.1) *The only two-cycle of K which is* $\neq 0$ *is* $\gamma^2 = \Sigma\sigma_i^2$.

Thus we already have for the Betti numbers

$$(18.2) \qquad R^0(K) = 1, \qquad R^2(K) = 1.$$

The reduction process rests primarily upon the elementary property:

*(18.3) *Let $\bar{E}_1^2, \bar{E}_2^2$ be two closed two-cells (of a complex) whose boundaries B_1, B_2 are such that $B_1 \cap B_2 = \bar{l}$, where l is an arc. Then $E^2 = E_1^2 \cup E_2^2 \cup l$ is a two-cell and $\bar{E}^2$ is a closed two-cell whose boundary $B = B_1 \cup B_2 - l$.*

19. Let us take for convenience the complex K in a geometric representation. The first step in the reduction consists in flattening out K upon a plane, much as one does with a box cut along some edges and flattened out on a table.

Take any triangle abc of K and correspondingly draw a triangle $a_1b_1c_1$ in a plane $\mathfrak{E}^2$. If bcd is adjacent to abc in K draw a triangle $b_1c_1d_1$ in the plane not overlapping abc. This is kept up until K is exhausted,

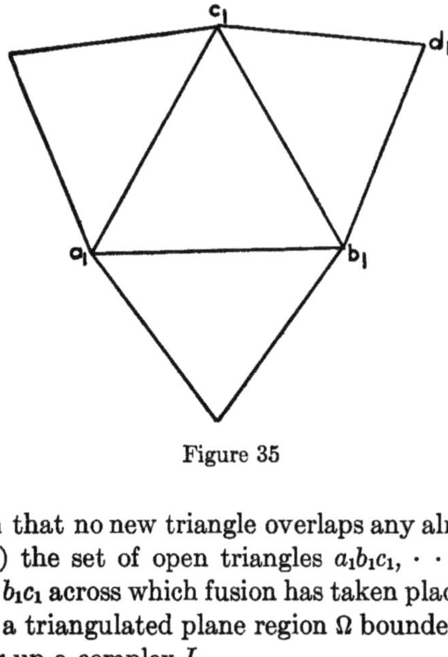

Figure 35

care being taken that no new triangle overlaps any already drawn. According to (18.3) the set of open triangles $a_1b_1c_1, \cdots$, together with the sides such as b_1c_1 across which fusion has taken place make up a two-cell E^2, which is a triangulated plane region Ω bounded by a polygon Π and thus making up a complex L.

During the fusion process we shall take care that two sides crossed consecutively should be sides of a triangle. Consider now the transformation t from the vertices of L to those of K whereby $ta_1 = a$, $tb_1 = b$, $\cdots$. Since t sends the vertices of a triangle of L into those of a triangle of K, t may be extended affinely over the whole set of triangles and thus becomes a mapping $|L| \rightarrow |K|$ still called t. Since each triangle of L goes into a single triangle of K, and each side or vertex in Ω into a single side or vertex of K, t is topological in Ω. Since each one-simplex of K is incident with two triangles, t sends a pair of the sides of the polygon Π into a single one-simplex of K. Thus topologically one may pass from the closed region $\bar{\Omega} = \Omega \cup \Pi$ to K by identifying the sides of Π in pairs. Then the images of the sides and vertices of Π under t make up a subcomplex G of K which is a linear graph. Since t maps Ω topologically on $|K| - |G|$, this last set is a two-cell and hence it is connected.

20. We shall now introduce the normalization process. In substance it is based upon a method due to Brahana (*Annals of Mathematics*, Vol. 23 (1921), pp. 144-168) with important simplifications communicated to the author by A. W. Tucker.

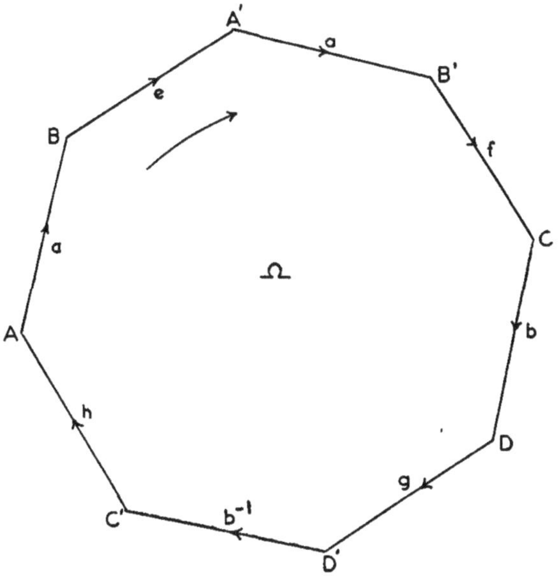

Figure 36

Consider then a plane convex region Ω bounded by a $2n$-sided polygon Π from which Φ may be obtained by identification of sides. Let $(AB, A'B'), (CD, C'D'), \cdots$, be identified pairs of sides and mark a positive sense on the boundary. We assume that $A = A', B = B', \cdots$. Let a designate both AB and $A'B'$ described from A to B, or from A' to B', and let a^{-1} denote them described oppositely. Similarly for CD, $C'D'$ and a label b, etc. Starting from one of the vertices say A, let Π be described once positively and write down as a product the symbols of the sides as they are passed in succession. We thus obtain the *surface symbols*. For example to Fig. 36 there corresponds the symbol $aeafbgb^{-1}h$.

We note that since the sides are identified in pairs, if a is a side the surface symbol contains one of the combinations $\cdots a \cdots a \cdots$, $\cdots a \cdots a^{-1} \cdots$, the symbol a occurring nowhere else. The combination of the first type is called *nonorientable*, and that of the second

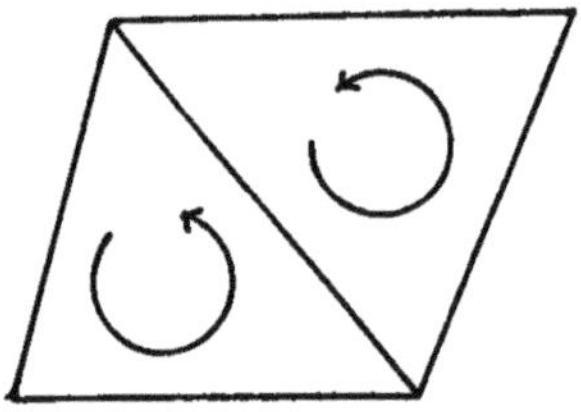

Figure 37

type is called *orientable*. If all the combinations are orientable Φ is said to be orientable, otherwise it is said to be nonorientable.

It may be pointed out, and this is a readily verified property, that orientability implies that the triangles of K may be "oriented" (in an

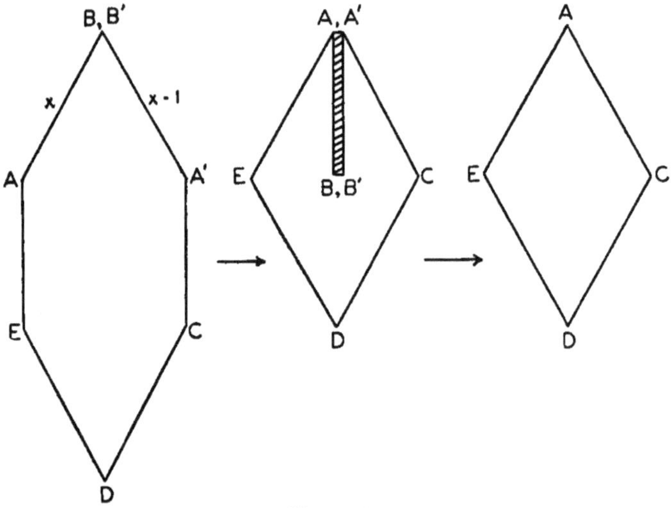

Figure 38

intuitive sense) so that adjacent triangles always have their orientations disposed as in Fig. 37. On the contrary if Φ is nonorientable the endeavor to carry this scheme through breaks down somewhere. Orientation will be discussed in full in the next chapter. As we shall base no argument on it in the present chapter the above remark is sufficient.

Returning now to the surface symbols, we shall describe several operations on the symbols which obviously do not modify the topology of Φ. Two symbols M,N derived from one another by our operations are said to be *equivalent*, written $M \simeq N$. Our operations are:

I. *Cyclic permutation of the factors.*

II. *Suppression or introduction of a pair of consecutive factors* x,x^{-1}.

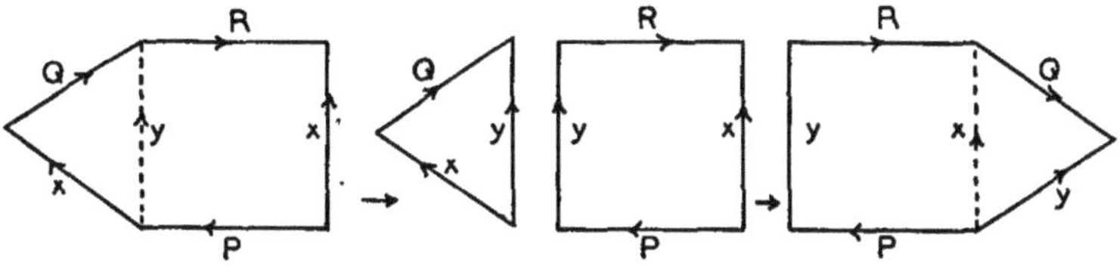

Figure 39

It is only necessary to justify suppression. Suppose we have $Px\,x^{-1}\,Q$ where here and later $P,Q, \cdots$ stand for certain products of sides. The modification is indicated in Fig. 38 which also shows its justification.

III. $Px\,Q\,Rx^{-1} \simeq Py\,RQ\,y^{-1}$ *where* $y = xQ,\ x^{-1} = Qy^{-1}$.

The passage from the first to the second and vice versa is justified by Fig. 39. The intermediate figure illustrates an operation of severing and pasting together which justifies the process.

IV. $PxQRx \simeq PyRy\,Q^{-1}$ *where* $y = xQ,\ x = yQ^{-1}$.

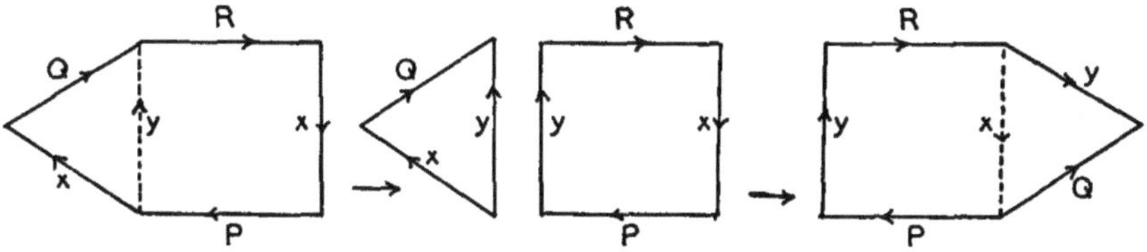

Figure 40

This operation is justified by Fig. 40.

Hence by repeated application $PxxQ \simeq yy\,PQ \simeq PQ\,yy$.

21. We are now ready for normalization.

Orientable symbols. Let Σ denote the symbol of Φ. For any factor x we have an orientable combination $\cdots x \cdots x^{-1} \cdots$. We will say that x,x^{-1} and y,y^{-1} are *separated pairs* if $\Sigma = Px\,Qy\,Rx^{-1}\,Sy^{-1}$ to within a cyclic permutation of the elements.

Suppose that Σ is the symbol just written. Repeated application of rules I and III yields

$$\Sigma \simeq Px_1\, yR\, Qx_1^{-1}\, S\, y^{-1} \simeq y_1x_1^{-1}\, SRQ\, y_1^{-1}\, Px_1$$

$$\simeq y_1x_2^{-1}\, y_1^{-1}\, PSRQ\, x_2 \simeq x_2\, y_1\, x_2^{-1}\, y_1^{-1}\, PSRQ.$$

So long as there remain separated pairs the above process may be applied. Since the number of terms is not changed in the process, repetition will ultimately lead Σ to the form HL where

$$H = a_1b_1\, a_1^{-1}\, b_1^{-1} \cdots a_pb_p\, a_p^{-1}\, b_p^{-1} \tag{21.1}$$

and L has no more separated pairs. If $L = xM$, then $L = x\, Nx^{-1}\, P$, where N contains no separated pairs. Hence $N = zQz^{-1}Q'$, where Q contains no separated pairs. Continuing thus we will find that $L \simeq Ruu^{-1}\, S \simeq RS$ by application of rule II. Thus the number of factors in L may be reduced by two. This may be kept up until L is completely suppressed and so finally $\Sigma \simeq H$. The reduced symbol (21.1) is said to be in the *normal form for the orientable case.* The number p in the reduced symbol (21.1) is known as the *genus* of the surface Φ.

The reduction just carried out assumes the existence of at least two separated pairs. If no such pair exists the initial symbol is of the type L. Instead of assuming the polygon in the plane let it be taken on a sphere S, or equivalently let the plane be closed at infinity by a point thus making it a sphere. The same argument for the suppression of L may be applied and will reduce the polygon Π to a point and Ω to the sphere. Under these conditions t is a homeomorphism and hence Φ is a (topological) sphere. It is natural to ascribe to the sphere the genus $p = 0$.

Nonorientable symbols. If $\Sigma = Px\, Qx\, R$, then $\Sigma \simeq Py\, y\, Q^{-1}\, R \simeq zz\, PQ^{-1}\, R$. By repeated application of rule IV, we may therefore reduce Σ to the form KL where

$$K = a_1a_1 \cdots a_qa_q, \qquad q > 0 \tag{21.2}$$

and L contains no pair $a \cdots a$. If L contains no separated pairs $x \cdots y \cdots x^{-1} \cdots y^{-1}$, then reasoning as above we may suppress it and so $\Sigma \simeq K$. We may still have $\Sigma \simeq K\, x\, x\, Py\, Qz\, Ry^{-1}\, Sz^{-1}\, T$. This time by rule IV

$$\Sigma \simeq K\, x_1\, R^{-1}\, z^{-1}\, Q^{-1}\, y^{-1}\, P^{-1}\, x_1y^{-1}\, S\, z^{-1}\, T$$

$$\simeq K\, x_1\, R^{-1}\, S^{-1}\, y\, x_1^{-1}\, P\, y\, Q\, z_1z_1\, T$$

$$\simeq K\, x_1\, R^{-1}\, S^{-1}\, P^{-1}\, x_1\, y_1\, y_1\, Q\, z_1\, z_1\, T$$

$$\simeq K\, x_2\, x_2\, P\, S\, R\, y_1\, y_1\, Q\, z_1\, z_1\, T$$

$$\simeq K\, x_2\, x_2\, y_2\, y_2\, z_2\, z_2\, PSRQT.$$

Thus the separated pair has been suppressed but the number of factors aa at the beginning has been augmented by two. Since the operation does not change the total number of factors, Σ will ultimately be reduced to the form KL where L has no separated pairs $x \cdots y \cdots x^{-1} \cdots y^{-1}$ and no pair $a \cdots a$. As before L may be suppressed, and Σ will again be reduced to the form K. This is the normal form for the non-orientable case.

(21.3) *In the two reduced forms all the vertices of the polygon* Π *correspond to the same point of the surface* Φ.

Consider the orientable form H of (21.1) and let the vertices of a_1 be A,B and let those of a_1^{-1} be A',B' where A' is identified with A and B' with B. Since B,B' are the vertices of b_1 they have the same image

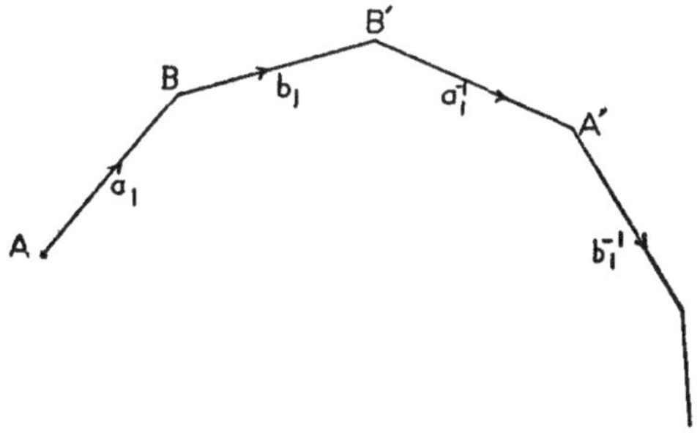

Figure 41

in Φ. If we start with the last vertex of b_1^{-1} and go in reverse direction we find that A',B' and hence A,B have the same image in Φ. Thus in each quadruple $a_i b_i a_i^{-1} b_i^{-1}$ the ends of the sides have the same image in Φ. Since consecutive quadruples have a common vertex (21.3) follows in the orientable case. The treatment of the nonorientable symbol (21.2) is essentially the same and is omitted.

22. Up to the present the two distinct normal forms, H and K, have not been associated with anything topological. As a matter of fact to do so in any thorough manner one must have recourse to integral chains and cycles and the related orientability theory which is fully developed in the subsequent chapters. There exists however a simple characterization of the two types based on a certain connectedness property. The informed reader will recognize its relation with the orientability theory. It may however be framed independently as we shall now see.

We first recall the configuration known as a *Möbius strip* which is obtained by taking a rectangle $ABCD$ and identifying the side AB with the side DC (Fig. 42) so that A goes into C and B into D. Let the construction be so carried out that the mid-points M,N of AB and CD are identified. As a consequence the line MN will go into a Jordan curve J on the strip. It is readily seen that J possesses the following property relative to the strip: Given any ϵ there exists a neighborhood $\mathfrak{N}(J)$ of J on the strip such that $\mathfrak{N}(J) - J$ is connected and all points of $\mathfrak{N}(J)$ are not farther than ϵ from J. A Jordan curve J in the surface Φ is said to be *one-sided* or *two-sided* in Φ accordingly as it does or does not possess the preceding property relative to Φ. We leave to the reader the proof of the following property:

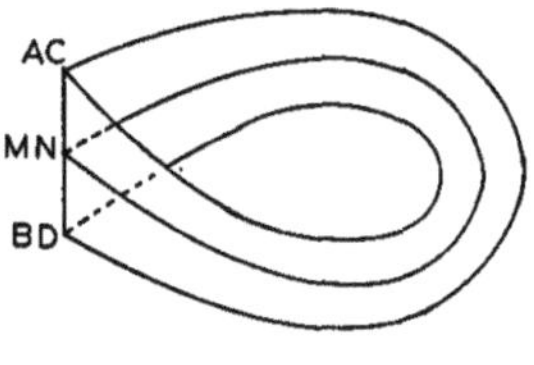

Figure 42

*(22.1) *A n.a.s.c. for a closed surface to be nonorientable [orientable] is that it contain some [no] one-sided Jordan curves.*

23. The Euler-Poincaré formula for a closed surface. We return to the situation of (19) with the complex K, the graph G and the mapping $t : |L| \to |K|$. The Euler-Poincaré relation for K is

$$\chi(K) = R^0(K) - R^1(K) + R^2(K). \tag{23.1}$$

In view of (18.2) this reduces here to

$$\chi(K) = 2 - R^1(K). \tag{23.2}$$

We first show that (23.2) is reducible to

$$R^1(K) = R^1(G). \tag{23.3}$$

In fact since $|G|$ is the image of Π under a mapping, and Π is connected, so is $|G|$. Hence $R^0(G) = 1$. Thus

$$\chi(G) = 1 - R^1(G). \tag{23.4}$$

On the other hand if at each stage of the fusion process one calculates χ as the expression $\alpha'^0 - a'^1 + a'^2$ where α'^2 is unity plus the number of remaining triangles (α'^2 is the number of two-cells of the configuration), α'_1 the number of sides not yet crossed during the fusion process and $\alpha'^0 = \alpha^0$, one finds that χ remains fixed. At the end its value is $1 + \chi(G)$. Hence

$$\chi(K) = 1 + \chi(G) = 2 - R^1(G)$$

and so if (23.3) holds, (23.2) and hence (23.1) will follow.

Proof of (23.3). Let γ^1 be any one-cycle of K. We may write $\gamma^1 = C^1 + D^1$, $C^1 \subset G$, $D^1 \subset K - G$. Owing to the one-one correspondence between the simplexes of $K - G$ and those of $L - \Pi$, there exists a chain D'^1 in $L - \Pi$ sent onto D^1 by the chain mapping $\tau: L \to K$ defined by t. Thus $D^1 = \tau D'^1$. Since $FD^1 = FC^1 \subset G$, FD'^1 is in Π. On the other hand the Kronecker index $\kappa(FD'^1) = 0$, (15.6), and therefore FD'^1 bounds in Π. Thus Π contains a chain C'^1 such that $FC'^1 = FD'^1$. Hence $C'^1 + D'^1$ is a one-cycle of L. Since $|L| = \bar{\Omega}$ is a closed two-cell, this one-cycle bounds. Hence L contains a chain C'^2 such that $FC'^2 = C'^1 + D'^1$. Applying τ we have then a chain $\tau C'^2 = C^2$ in K such that $FC^2 = \tau(C'^1 + D'^1) = \tau C'^1 + D^1 = \tau C'^1 + C^1 + \gamma^1$. Thus in K: $\gamma^1 \sim \tau C'^1 + C^1$ which is in G. In other words:

(23.5) *Every one-cycle of K is homologous to a cycle of G.*

Let now $\gamma_1^1, \cdots, \gamma_r^1$, $r = R^1(G)$, be independent cycles of G. If they were dependent in K there would exist a relation

$$FC^2 = \Sigma g_i \gamma_i^1, \; C^2 \subset K,$$

with the g_i not all zero mod 2. Since $|K| - |G|$ is a two-cell, it is connected. As a consequence we may reason as for (11.5), and show that since $C^2 \not\approx 0$, it is the sum γ^2 of all the σ_i^2. Thus $C^2 = \gamma^2$, $FC^2 = 0$. Therefore the γ_i^1 are independent in K. By (23.5) every one-cycle of K depends on the γ_i^1 and so $R^1(K) = r = R^1(G)$. This proves (23.3), hence also the Euler-Poincaré relation in the case considered.

Since $\chi(K)$ depends solely upon $|K|$, (15.5), it is referred to henceforth as the characteristic of Φ, written $\chi(\Phi)$.

Application to the closed two-cell. Let $\bar{E}^2$ be a closed two-cell whose boundary is J. Let K be a complex made up of the elements of a triangulation of $\bar{E}^2$ with $J = |L|$, L a subcomplex of K. Identify $\bar{E}^2$ with the upper hemisphere of a sphere S, and J with the equator. The symmetric of the elements of K as to the equatorial plane make up a complex K_1 isomorphic with K and $K \cup K_1$ is a complex made up of the elements of a triangulation of S. Hence

$$\chi(S) = 2 = \chi(K) + \chi(K_1) - \chi(J) = 2\chi(K).$$

Therefore $\chi(K) = 1$. This number is again independent of K. It is by definition the characteristic of $\bar{E}^2$ and is written $\chi(\bar{E}^2)$. Notice that $\chi(K - L) = \chi(K) - \chi(L) = \chi(K)$. Thus for the same reason we may introduce $\chi(E^2)$ and its value is still unity.

24. Characterization of orientable surfaces. Consider the normal form H, of (21.1). The corresponding polygon Π and region Ω together have the characteristic unity of the closed triangle, or $\chi(\bar{\Omega}) = 1$. Under the mapping $t : \bar{\Omega} \to \Phi$, the $4p$ sides reduce to $2p$ and the $4p$ vertices to one. Hence

$$\chi(\Phi) = 1 - 2p + 1 = 2 - 2p = 2 - R^1,$$

and so $R^1 = 2p$. Therefore:

(24.1) **Theorem.** *The Betti number $R^1(\Phi)$ mod 2 of a closed orientable surface is even and equal to twice the genus. Hence the genus is a topological invariant.*

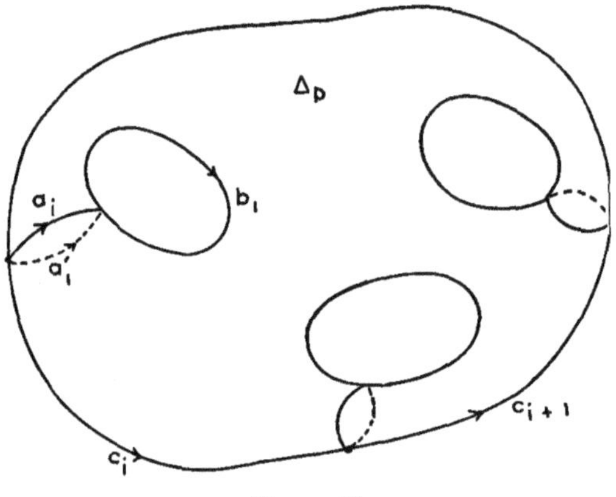

Figure 43

Two orientable connected closed surfaces are homeomorphic if they have the same genus p since they have then identical reduced normal forms. On the other hand since the genus has topological character, two closed connected orientable surfaces with different values of the genus cannot be homeomorphic. Therefore

(24.2) **Theorem.** *A closed orientable surface is completely characterized topologically by its genus.*

Consider a two-sided disk Δ_p with p holes. In Fig. 43 we have taken $p = 3$ and we assume $c_{i+3} = c_i$, etc. Under the circumstances the two-cell left from the top of the disk with the arcs a_i, b_i, c_i removed has a symbol $a_1b_1a_1^{-1}c_1a_2 \cdots a_p^{-1}c_p$, while the bottom two-cell has the symbol $a'_1b_1a'^{-1}_1c_1a'_2 \cdots a'^{-1}_p c_p$. If they are matched together along c_p the symbol for the resulting polygon is $a_1b_1a_1^{-1}a_2 \cdots a_p^{-1}a'_pb_p^{-1}a'^{-1}_p \cdots a'^{-1}_1$. There are the p separated pairs $a_i \cdots b_i \cdots a_i^{-1} \cdots b_i^{-1}$. Hence the surface symbol is that of an orientable surface of genus p. Therefore every surface of genus $p > 0$ is homeomorphic to a two-sided disk with p holes.

25. Characterization of nonorientable surfaces. Consider the normal form K of (21.2). This time the passage from $\bar{\Omega}$ to Φ replaces the polygon Π of $2q$ sides and $2q$ vertices by q arcs and one vertex. Hence $\chi(\Phi) = 1 - q + 1 = 2 - q$, and so $R^1 = q$. Here R^1 may have any positive value and we have:

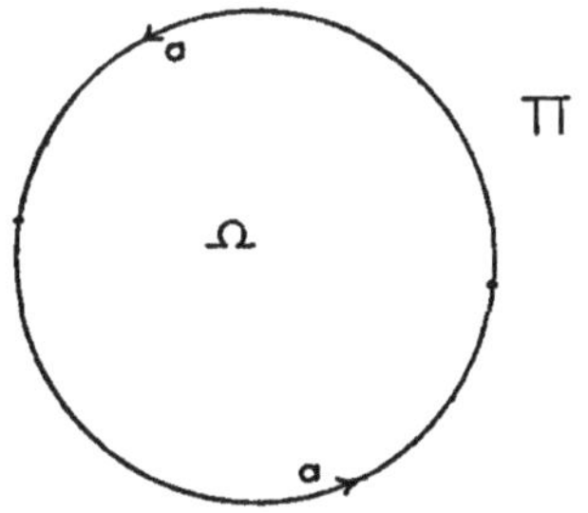

Figure 44

(25.1) **Theorem.** *A closed nonorientable surface is completely characterized topologically by its Betti number* R^1.

Combining with (24.2) we have then:

(25.2) **Theorem.** *A closed surface* Φ *is completely characterized by the value of its Betti number* R^1 *and by its orientability or nonorientability. When* R^1 *is odd* Φ *is necessarily nonorientable.*

(25.3) *Projective plane.* Topologically this surface is characterized as a closed nonorientable surface whose Betti number R^1 is unity. One may choose as the corresponding pair Ω,Π associated with the surface symbol aa, a circular region whose boundary is divided in two by a diameter (Fig. 44). The surface is obtained from the closed region by identifying the two ends of each diameter.

General nonorientable case. From a sphere S remove q disjoint circular regions and let $C_1, \cdots, C_q$ be their boundaries. A Möbius strip has for boundary a Jordan curve. Take q such strips $M_1, \cdots, M_q$ and identify the boundary of M_i with C_i. The result is a closed surface Φ

which is nonorientable since the median line of M_i is one-sided in M_i and hence in Φ. Let us calculate $\chi(\Phi)$. For M_i it is the same as for a rectangle with two opposite sides identified, or $\chi(M_i) = 1 + 1 - 2 = 0$. The removal of q two-cells reduces $\chi(S)$ by q units. Hence

$$\chi(\Phi) = \chi(S) - q + \Sigma\chi(M_i) - \Sigma\chi(C_i) = 2 - q.$$

Therefore every closed nonorientable surface whose Betti number $R^1 = q$ is homeomorphic to the model just constructed.

PROBLEMS

1. A *figure 8 curve* is the homeomorph of the figure formed by two tangent circles. Show that a figure 8 curve drawn in the two-sphere S divides S into three regions.

2. Let the graph G be a tree. If the set $|G|$ is in the two-sphere S then it does not disconnect S.

3. Given a closed subset A of the two-sphere S show that the number ρ of components of $S - A$ is a topological invariant of A and hence does not depend upon its imbedding in S.

4. Prove the *Jordan-Brouwer theorem:* an $(n - 1)$-sphere S^{n-1} contained in S^n subdivides S^n into two regions whose common boundary is S^{n-1}.

5. Prove that a torus T in S^3 subdivides S^3 into two regions with the common boundary T.

6. Let J be a Jordan curve in S^3. Prove that $S^3 - J$ is connected. Similarly if J is replaced by a figure 8 curve H. Is it possible to differentiate between $S^3 - J$ and $S^3 - H$ by a topological property of the curves J, H?

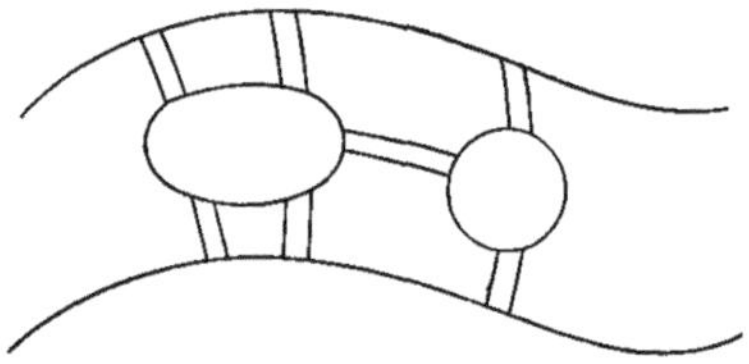

Figure 45

7. *The problem of the seven bridges of Koenigsberg:* The plan of the city of Koenigsberg is as indicated in Fig. 45 and there are seven bridges. Is it or is it not possible to walk from one point and back to the same point in the city crossing each bridge once and only once. (The solution was given by Euler.)

8. Given in a city a gas plant G, an electric plant E and a water plant W, is it possible to draw the appropriate mains to three residences A,B,C so that none passes under another?

9. A closed nonorientable surface may be represented by one of the surface symbols

$$a_1b_1a_1^{-1}b_1^{-1} \cdots a_pb_pa_p^{-1}b_p^{-1}\, a_{p+1}\, a_{p+1}$$

$$a_1b_1a_1^{-1}b_1^{-1} \cdots a_{p-1}b_{p-1}a_{p-1}^{-1}\, b_{p-1}^{-1}\, a_p\, b_p\, a_p^{-1}\, b_p;$$

depending upon whether its characteristic is odd or even (A. W. Tucker).

10. A closed orientable surface of genus p may be represented by one of the surface symbols (A. W. Tucker):

$$a_1a_2 \cdots a_{2p}\, a_1^{-1}\, a_2^{-1} \cdots a_{2p}^{-1}$$

$$a_1 \cdots a_{2p+1}\, a_1^{-1} \cdots a_{2p+1}^{-1}.$$

III. Theory of Complexes

The questions discussed in the preceding chapter for low dimensional complexes are now to be treated for all complexes. In the present chapter the basic definitions are given, more general chains and related concepts are introduced and algebraic questions come more strongly to the fore. However we do not go beyond homology questions and some elementary applications. Transformations of complexes and related topics are reserved for the following chapters.

§1. Intuitive Approach.

1. A good guide in the type of questions that concern us is always provided by integration. We have seen in the *Survey* (6) that the study

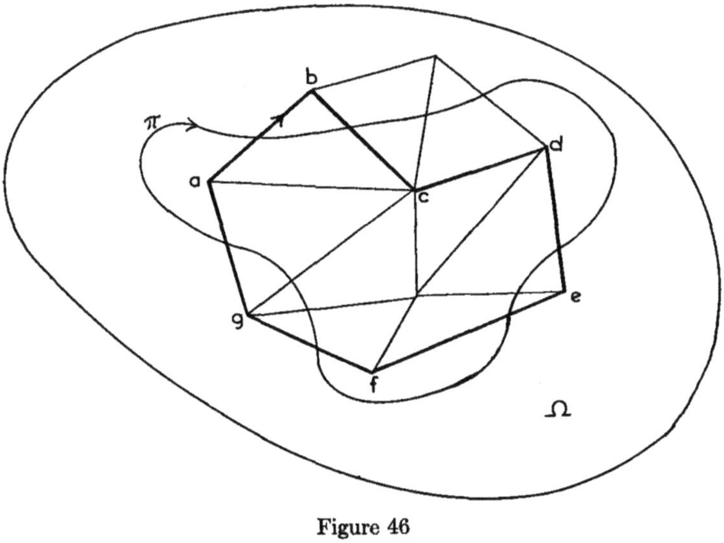

Figure 46

of the determinations of a simple integral $J(\pi)$ leads to a sort of weak homotopy where cancellation of opposite paths is permitted. There still remain a large number of paths to be considered and some method for reducing their number is most desirable. Suppose that the integral J is taken over a closed triangulated region Ω. A closed path π in the region is then deformed into a path $\pi_1 = abcdefga$ made up of oriented sides of the triangulation and we have $J(\pi) = J(\pi_1)$. The path π_1 is naturally described by the one-chain $ab + bc + \cdots + ga$. One might even come upon a path such as $abca$ repeated twice and whose proper description would be as a one-chain $2ab + 2bc + 2ca$ with integral coefficients. In the light of the discussion in the *Survey*, the closed paths would correspond to the cycles, the "negligible" paths to the multiples of triangle boundaries and we could proceed from here in setting up the homology groups. To give the scheme significance in relation to triangulation one would have to show that these groups are independent of the triangulation by means of which they are defined. For higher dimensional integrals the situation is more or less the same.

These considerations point therefore to the following program: To set up a suitable n-dimensional homology theory for complexes and to prove that the theory has topological character. Once this theory is properly built up it will be found to have many applications besides integration. In fact this has already been clearly shown for the two-dimensional case in the preceding chapter.

§2. Simplexes and Simplicial Complexes.

2. Our first step is to define simplexes, the analogues of triangles, then put them together so as to make up the complexes.

A *p-dimensional simplex,* or more simply a *p-simplex* is just a set of $p + 1$ elements $a_0, \cdots, a_p$. The p-simplex is often written σ^p or σ, also ζ^p or ζ. The a_i are called the *vertices* of the simplex. The number p is the *dimension* of σ^p, written also dim σ. A subset σ^q of σ^p is called a *q-face* of σ^p, a *proper* q-face when $q < p$. The vertices are the zero-faces and σ^p is its own unique p-face. We also say: σ^p and σ^q are *incident* and denote their relationship symbolically by an ordering relation $\sigma^q < \sigma^p$ or $\sigma^p > \sigma^q$.

The simplex σ^p with the vertices $a_0, \cdots, a_p$ is often written $a_0 \cdots a_p$. If $a^p = a_0 \cdots a_p$ and $\sigma^q = b_0 \cdots b_q$ then the simplex $a_0 \cdots a_p b_0 \cdots b_q$ is written $\sigma^p\sigma^q$.

Orientation. Let $\sigma^p = a_0 \cdots a_p$ be considered as a skew-symmetric symbol. A selection of one of the two possible orders is called *orienting* σ^p.

Incidence numbers. Suppose all the simplexes oriented and let us introduce between the p-simplexes and $(p - 1)$-simplexes a number $[\sigma^p : \sigma^{p-1}]$ called their *incidence number* and defined as follows: If $\sigma^p = \epsilon a\sigma^{p-1}$, $\epsilon = \pm 1$, then $[\sigma^p : \sigma^{p-1}] = \epsilon$; otherwise $[\sigma^p : \sigma^{p-1}] = 0$. Thus when σ^p, σ^{p-1} are not incident their incidence number is zero.

If σ^{p-1} is a face of σ^p we shall say that they are *positively* [*negatively*] *related* whenever their incidence number is $+1$ [-1].

Example. If $\sigma^2 = abc$ (triangle) then bc is positively related to abc and negatively related to bac. Similarly bc is positively related to c and negatively to b.

3. The simplicial complex. A *simplicial complex* is a finite collection $K = \{\sigma\}$ of oriented simplexes such that if σ is in K so are all its faces. The *dimension* of K, written dim K, is the largest dimension of any simplex of K.

It is to be understood that if some or all the simplexes σ are reoriented then K is unchanged. The essential point is that *some* orientation must be assigned to them, but the specific choice made is immaterial. This means also that although some of the concepts given below are framed in terms of specific orientations of the simplexes, the theorems must be shown to be independent of these orientations.

One may refer to the above complexes as *oriented.* Whenever every σ is replaced by $\alpha(\sigma)\sigma$, $\alpha(\sigma) = \pm 1$, the complex K is said to be *reoriented.* We call $\alpha(\sigma)$ an *orientation function.*

The incidence numbers have the following fundamental property:

$$\sum_i [\sigma^p : \sigma_i^{p-1}][\sigma_i^{p-1} : \sigma^{p-2}] = 0. \tag{3.1}$$

The proof is quite simple. The product in the sum is zero unless σ_i^{p-1} is incident with σ^p and σ^{p-2}, so that σ^p is $\pm\, ab\sigma^{p-2}$, and at the same time $\pm\,\sigma_i^{p-1}$ is either $a\sigma^{p-2}$ or $b\sigma^{p-2}$. Since reorientation visibly does not affect the result we may assume that

$$\sigma^p = ab\sigma^{p-2}, \qquad \sigma_i^{p-1} = a\sigma^{p-2} \text{ or } b\sigma^{p-2}.$$

Writing now σ for σ^{p-2} the sum reduces to $[ab\sigma : a\sigma][a\sigma : \sigma] + [ab\sigma : b\sigma]$ $[b\sigma : \sigma] = 0$, since $[a\sigma : \sigma] = [b\sigma : \sigma] = +1$ and $[ab\sigma : a\sigma] = -[ab\sigma : b\sigma] = -1$.

By a *subcomplex* L of K is meant a subset of the set of the simplexes of K which is also a simplicial complex. That is to say, L consists of a set of simplexes of K such that if $\sigma \in L$ then every face of σ is likewise in L. The *p-section* K^p of K is the set of all the simplexes of K whose dimension $\leq p$. It is manifestly a subcomplex of K.

Example. σ^n together with all its faces make up a simplicial n-complex called the *closure* of σ^n, written $Cl\, \sigma^n$. The $(n-1)$-section of $Cl\, \sigma^n$ or set of all the proper faces of σ^n is an $(n-1)$-complex called the *boundary* of σ^n, written $\mathfrak{B}\, \sigma^n$.

Isomorphism. Two simplicial complexes K, K' are said to be *iso-*

morphic if their elements are in a one-one correspondence preserving their dimensions and incidences. Whenever K,K' are isomorphic it is possible to label the simplexes σ of K and σ' of K' so that σ_i^p, $\sigma_i'^p$ correspond to one another. Also if a_i, a_i' are their vertices the correspondence associates a_i with a_i' in such a way that if $a_i \cdots a_j \in K$ then $a_i' \cdots a_j' \in K'$. Except for the labels, then, K and K' are identical. It is to be pointed out however that $[\sigma_1 : \sigma_2]$ may differ from $[\sigma_1' : \sigma_2']$ by a factor $\alpha(\sigma_1')\alpha(\sigma_2')$ resulting from a reorientation.

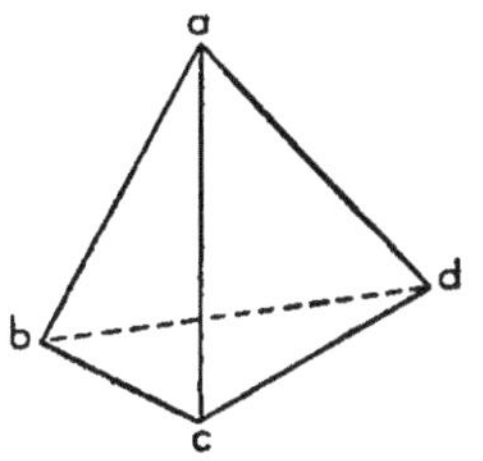

Figure 47

Star of a simplex. This is the set of all the simplexes having σ for a face and is written $St\,\sigma$.

Example. In Fig. 47, $K = \mathfrak{B}\,\sigma^3$, $\sigma^3 = abcd$. $Cl(ac)$ consists of ac, a, c while $St(ac)$ consists of ac, bac, dac.

§3. Chains, Cycles, Homology Groups.

4. Let $K = \{\sigma\}$ be an oriented simplicial complex. An *integral p-chain* of K is a linear form with integral coefficients

$$C^p = \Sigma g_i \sigma_i^p. \tag{4.1}$$

By addition, in the natural way, the integral *p-chains* make up a group $\mathfrak{C}^p$.

Example. In the complex represented in Fig. 48 the chain $C^1 = 2\,ab$

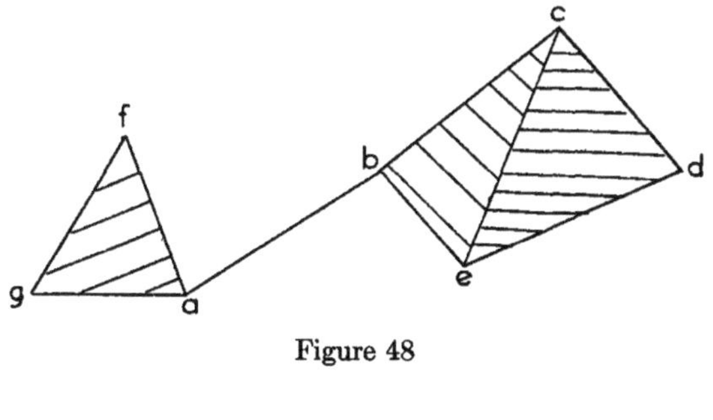

Figure 48

$+ 3\, bc - cd$ is an integral one-chain while $C^2 = afg - 4\, ced$ is an integral two-chain.

Conventions. I. If σ^p is replaced by its opposite $(-\sigma^p)$ then $g\sigma^p$ is to be replaced by $-g(\sigma^p)$.

II. If $g_i = gg'_i$, $C'_p = \Sigma g'_i \sigma^p_i$ then we write $C^p = gC'^p$.

III. Suppose that every element in (4.1) may be written $\sigma^q \sigma_i^{p-q-1}$, where σ^q is fixed. Then we also write $C^p = \sigma^q C^{p-q-1}$, $C^{p-q-1} = \Sigma g_i \sigma_i^{p-q-1}$.

IV. Under certain transformations occurring notably in Chapter IV, there may arise (sometimes tacitly) "*degenerate*" *simplexes*, i.e. with repeated vertices. Such a simplex, considered as a chain is, by definition, to be set equal to zero. Thus $2aab = 0$ as a two-chain.

V. Instead of "C^p is a chain of K" we shall sometime say: "C^p is in K," and write accordingly: $C^p \subset K$.

5. Define now as the *boundary* of σ^p_i the $(p - 1)$-chain

$$F\sigma^p_i = \Sigma[\sigma^p_i : \sigma_j^{p-1}]\sigma_j^{p-1}, \tag{5.1$_p$}$$

$$F\sigma^\circ_i = 0. \tag{5.1$_o$}$$

It is essentially the sum of the $(p - 1)$-faces of σ^p_i each affected with $+1$ or -1 accordingly as it is positively or negatively related to σ^p_i. Notice in particular that

$$Fa_o \cdots a_p = \Sigma(-1)^q\, a_o \cdots a_{q-1}\, a_{q+1} \cdots a_p,\ p > 0 \tag{5.2$_p$}$$

$$Fa_o = 0. \tag{5.2$_o$}$$

A boundary for C^p, written also FC^p, is defined by linear extension as

$$FC^p = \Sigma g_i[\sigma^p_i : \sigma_j^{p-1}]\sigma_j^{p-1}, \qquad p > 0 \tag{5.3$_p$}$$

$$FC^\circ = 0. \tag{5.3$_o$}$$

As an example, let again $C^1 = ab + bc + cd$ (Fig. 48). Then $Fab = b - a$, $Fbc = c - b$, $Fcd = d - c$. Hence $FC^1 = d - a$. Similarly we find $Fbce = bc + ce + eb$, $Fecd = ec + cd + de$ and consequently $Fbce + Fecd = F(bce + ecd) = eb + bc + cd + de$. These results have an obvious connotation.

From (5.1) and (3.1) we obtain for the chain C^p of (4.1):

$$FFC^p = \Sigma g_i[\sigma^p_i : \sigma_j^{p-1}][\sigma_j^{p-1} : \sigma_k^{p-2}]\sigma_k^{p-2} = 0. \tag{5.4}$$

or in operator form

$$FF = 0. \tag{5.5}$$

This relation is of fundamental importance in everything that follows. It may also be expressed as: *the operator F is of order two*. It is clear that:

(5.6) *The property (3.1) of the incidence numbers and the relation (5.5) for F are equivalent.*

An *integral p-cycle* γ^p is an integral p-chain such that

$$F\gamma^p = 0. \tag{5.7}$$

If γ^p, γ'^p are integral p-cycles, so is $\gamma^p - \gamma'^p$. Hence the collection $\mathfrak{Z}^p = \{\gamma^p\}$ of all the integral p-cycles is a subgroup of $\mathfrak{C}^p$.

We may now interpret (5.4), hence (5.5) as meaning that:

(5.8) *The boundary of a chain is a cycle.*

Chains such as FC^{p+1} are known as *bounding chains* or *bounding cycles*. Since $FC^{p+1} - FC'^{p+1} = F(C^{p+1} - C'^{p+1})$, the collection $\mathfrak{F}^p = \{FC^{p+1}\}$ is a subgroup of $\mathfrak{Z}^p$.

6. Since $\mathfrak{F}^p \subset \mathfrak{Z}^p$ and all the groups are commutative, we may introduce the factor group $\mathfrak{H}^p = \mathfrak{Z}^p/\mathfrak{F}^p$. This is the *integral p-th homology group* of the complex K. It may be viewed as the group of the p-cycles reduced modulo the bounding cycles, or of the p-cycles identified whenever they merely differ by a bounding cycle.

The "*homology*" terminology originated as follows. If the chains C_i^p are such that there can be found a C^{p+1} satisfying a relation

$$FC^{p+1} = \Sigma g_i C_i^p$$

then Poincaré expressed the fact by a relation which he described as a homology,

$$\Sigma g_i C_i^p \sim 0. \tag{6.1}$$

These relations may be combined linearly like ordinary linear equations after multiplication by integers. This becomes evident also if one writes (6.1) in the form

$$\Sigma g_i C_i^p = 0 \bmod \mathfrak{F}^p,$$

meaning that the left hand side is in $\mathfrak{F}^p$. The meaning of $C^p \sim D^p$ is then clearly: $C^p - D^p \in \mathfrak{F}^p$.

If the cycles $\gamma_1^p, \cdots, \gamma_r^p$ are such that there do not exist coefficients g_i not all zero such that $\Sigma g_i \gamma_i^p \sim 0$, that is $\Sigma g_i \gamma_i^p = 0 \bmod \mathfrak{F}^p$, then we say that the γ_i^p are *linearly independent with respect to homology*, or more briefly, *linearly independent*.

Since $\sim$ is manifestly an equivalence relation, it separates $\mathfrak{Z}^p$ into classes. These classes, known as *integral homology classes*, are merely the elements of $\mathfrak{H}^p$. The homology class of a cycle γ^p is the set of all cycles $\gamma^p + FC^{p+1}$, the coset of γ^p mod $\mathfrak{F}^p$ in $\mathfrak{Z}^p$.

Kronecker index of zero-chains. If $C^o = \Sigma g_i a_i$, where $\{a_i\}$ are the vertices of K, its *Kronecker index*, written $KI(C^o)$, is by definition Σg_i.

The index is manifestly a linear function. Its value arises chiefly from the property:

(6.2) *If* $C^o \sim 0$ *then* $KI(C^o) = 0$.

For if $\sigma' = a - b$ then $KI(F\sigma') = 1 - 1 = 0$. Since KI is linear every $KI(FC^1) = 0$ which is (6.2).

Effect of reorientation. Our groups would have but little value if their relations were affected by reorientation. Under the convention that $g(-\sigma) = (-g)\sigma$, if K is subjected to the orientation function $\alpha(\sigma)$, $g\sigma$ will go into $(\alpha(\sigma)g)\sigma$. Hence if $F\sigma = \Sigma[\sigma : \sigma']\sigma'$ then $F\sigma$ goes into $\Sigma\alpha(\sigma)\alpha(\sigma')[\sigma : \sigma']\sigma'$. The effect of the operation on $\mathfrak{C}^p$ is an isomorphism τ sending $\mathfrak{F}$, $\mathfrak{Z}$ into $\tau\mathfrak{F}$, $\tau\mathfrak{Z}$ and hence $\mathfrak{H}$ into an isomorph.

7. (7.1) *The homology groups for dimensions* $p < n - 1$ *are the same for* K *and its section* K^{n-1}.

For the groups in question are determined solely by means of the σ_i^q, $q \leqq n - 1$, which are the same for both complexes.

(7.2) *If* $\dim K = n$ *the corresponding groups* $\mathfrak{H}^n$ *and* $\mathfrak{Z}^n$ *are isomorphic and all the groups for dimension* $> n$ *are zero.*

8. Extension to general groups of coefficients. The only property of the integers utilized in the preceding algebraic developments is their forming an abelian group. Therefore nothing is changed if the g_i are allowed to be elements of any abelian group G. Functional designations: $\mathfrak{C}^p(K;G)$, $\mathfrak{C}^p(G)$, $\cdots$, are in order and have an obvious meaning.

In point of fact we shall confine G to one of the following groups: $\mathfrak{J}$ the group of the integers, $\mathfrak{J}_\pi$ the group of their residues modulo a prime π, $\mathfrak{R}$ the rational group. All but the first are *fields*. For convenience we shall also refer to $\mathfrak{R}$ as "group of residues mod 0." Instead of "chains, $\cdots$, over $\mathfrak{J}$, $\mathfrak{J}_\pi$, $\mathfrak{R}$ "we shall say "integral chains, $\cdots$, chains mod π, $\cdots$, rational chains, $\cdots$," with designations such as: $\mathfrak{C}^p$, $\mathfrak{C}^p_\pi$, $\mathfrak{C}^p_0$, $\cdots$.

When G is a field the groups $\mathfrak{C}, \mathfrak{Z}, \mathfrak{F}$ become vector spaces over G. The standard convention regarding factor spaces is made regarding the homology groups: if γ^p is a cycle and Γ^p its homology class then $g\gamma^p$ is a cycle whose homology class is written $g\Gamma^p$. This convention makes likewise $\mathfrak{H}^p$ a vector space over G.

Wherever homomorphisms of groups $\mathfrak{C}$, $\cdots$ over a field G into other such groups occur in the sequel, it will be very easy to verify that they are linear transformations of the vector spaces. This verification will generally be omitted and it is to be understood that in the cases referred to one may read "linear transformation" for "homomorphism," and "nonsingular linear transformation" for "isomorphism."

9. When G is a field the dimension of the homology vector space $\mathfrak{H}^p(K;G)$ is known as the p-th *Betti number* of K over G, written $R^p(K,G)$, or $R^p_\pi(K)$ for $G = \mathfrak{J}_\pi$.

It may be stated (without proof) that when G is a field of characteristic π then $R^p(K;G) = R^p_\pi(K)$. Thus by confining our attention to the groups $\mathfrak{J}_\pi$ we obtain actually all the Betti numbers.

One may think of $R^p(K;G)$ as the maximum number of p-cycles over G which are linearly independent of the bounding cycles of the same kind. In that sense one may also consider the maximum number of integral p-cycles linearly independent of the bounding integral p-cycles. By a well known argument this number is the same as for the rational cycles, that is to say it is in fact R^p_o.

The above remarks establish the clear connection with the Betti numbers as conceived in connection with integration.

From (7.2) follows:

(9.1) *R^n_π is the maximum number of linearly independent n-cycles* mod π, *i.e. the dimension of* $\mathfrak{Z}^n_\pi$.

10. Cyclic terminology. If there exists an integral p-cycle $\gamma^p \not\sim 0$ such that every cycle over any $G \sim g\gamma^p$, $g \in G$, then K is said to be *cyclic in the dimension p*. If all the p-cycles are ~ 0, (the p-th homology groups are zero), then K is said to be *acyclic in the dimension p*. The complex is said to be *p-cyclic* when it is cyclic in the dimension p and acyclic in the rest. Finally K is said to be *acyclic* when it is acyclic in all dimensions. (This last statement is really purely formal since simplicial complexes are never acyclic. However the same terminology will be applicable to a much broader situation and hence it is given here in a manner requiring no repetition later.)

(10.1) *Example.* A zero-simplex σ^o is zero-cyclic.

(10.2) When K is cyclic in the dimension p its Betti numbers R^p_π are all unity. When it is acyclic in the dimension p all the R^p_π are zero.

11. Homology groups of some elementary complexes. Let us first make an elementary observation. If $\{a_i\}$ are the vertices of K and any a_i may be joined to one of them say a_1 by a sequence $a_1, a_1', \cdots, a_r', a_i$, where consecutive vertices are joined by a $\sigma^1 \in K$, then K is cyclic in the dimension zero. In fact we have $a_1 \sim a_1' \sim \cdots \sim a_r' \sim a_i$. Hence if $C^o = \Sigma g_i a_i$ is any zero-cycle then

$$C^o \sim (\Sigma g_i)a_1 = KI(C^o)a_1.$$

Thus every C^o is related to a_1 by a homology. On the other hand $a_1 \not\sim 0$ since its index $\neq 0$. This is sufficient to prove our assertion.

Joins. If $K = \{\sigma\}$ is any collection of simplexes the collection of simplexes $\{a\sigma\}$, where a is not a vertex of any σ, is written aK and called the *join* of a and K. If K is a simplicial complex $a \cup aK \cup K = L$ is likewise a simplicial complex and is known as the *closed join* of a and K.

It is to be recalled that under our conventions when $C^p = \Sigma g_i \sigma_i^p$, then the chain $\Sigma g_i\, a\sigma_i^p$ is written aC^p. Moreover when a is a vertex of σ_i^p then $(a\sigma_i^p) = 0$ as a chain.

The boundary relation for $a\sigma_i^p$ yields at once

$$Fa\sigma_i^p = \sigma_i^p - aF\sigma_i^p,$$

a relation which holds even when a is a vertex of σ_i^p. As a consequence

$$FaC^p = C^p - aFC^p. \tag{11.1}$$

We will now prove

(11.2) *If K is a simplicial complex then the closed join $L = a \cup aK \cup K$ is zero-cyclic.*

Since every vertex a_i of L is either a or else joined to it by the one-simplex aa_i, L is cyclic in the dimension zero. It is therefore sufficient to show that a cycle $\gamma^p \sim 0$ for $p > 0$. Now $\gamma^p = aC^{p-1} + C^p$, C^{p-1} and C^p in K. By (11.1): $F\gamma^p = 0 = C^{p-1} + FC^p - aFC^{p-1}$. Since the elements in the last term have the vertex a, and those in the first two do not, we must have $FC^{p-1} = 0$, $C^{p-1} + FC^p = 0$. Therefore $\gamma^p = FaC^p \sim 0$. This proves (11.2).

(11.3) **Corollary.** *$Cl\ \sigma^n$ is zero-cyclic.*

When $n = 0$ this is obvious. When $n > 0$ then $\sigma^n = a\,\sigma^{n-1}$ and $Cl\ \sigma^n$ is the closed join of a and $Cl\ \sigma^{n-1}$. Hence it is zero-cyclic.

(11.4) **Corollary.** *$\mathfrak{B}\sigma^{n+1}$, $n > 0$, is cyclic in the dimensions 0,n and acyclic in the rest.*

Since $\mathfrak{B}\sigma^{n+1}$ is the n-section of $Cl\ \sigma^{n+1}$ they have the same homology groups for dimensions $< n$. Hence $\mathfrak{B}\sigma^{n+1}$, like $Cl\ \sigma^{n+1}$, is cyclic in the dimension zero, and acyclic in every dimension p for $0 < p < n$. There remains then to prove it cyclic in the dimension n. Now for that dimension the properties "~ 0" and "$= 0$" are equivalent (7.2). Thus we merely need to examine the n-cycles relatively to (ordinary) linear dependence (without reference to bounding). When $\sigma^{n+1} = a_o \cdots a_{n+1}$, every n-cycle of $\mathfrak{B}\sigma^{n+1}$ is of the form

$$\gamma^n = \Sigma(-1)^i g_i\, a_o \cdots a_{i-1}\, a_{i+1} \cdots a_{n+1}.$$

By direct computation we find

$$F\gamma^n = \sum_{i<j}(-1)^{i-j}(g_i - g_j)\, a_o \cdots a_{i-1}\, a_{i+1} \cdots a_{j-1}\, a_{j+1} \cdots a_{n+1}$$

which is identically zero if and only if $g_i = g_j$. Therefore $\gamma^n = gF\sigma^{n+1}$. Thus $\mathfrak{B}\, \sigma^n$ is cyclic in the dimension n and (11.4) folllows.

§4. Geometric Complexes.

12. After the preceding algebraic considerations we return to geometry. Our primary objective is to retrace our steps from simplicial complexes back to polyhedra. This will make the connection between the two quite obvious. As we shall make an intensive use of barycentric coordinates a few preliminary remarks concerning them are in order.

Barycentric coordinates. Let $\mathfrak{E}^n$ be an Euclidean space referred to coordinates $x_1, \cdots, x_n$. A set of $p + 1$ points $a_i = (a_{i1}, \cdots, a_{in})$, $i = 0, \cdots, p$ of $\mathfrak{E}^n$ is said to be *linearly independent* if it spans an $\mathfrak{E}^p$, i.e. if the matrix $||a_{i1}, \cdots, a_{in}, 1||$ is of rank $p + 1$. The points of $\mathfrak{E}^p$ may then be represented as

$$(12.1) \qquad x = \Sigma t_i a_i, \qquad \Sigma t_i = 1,$$

meaning that x has the coordinates $\Sigma t_i a_{ih}$. The t_i are *barycentric coordinates* for $\mathfrak{E}^p$, referred to the points a_i. As an example the $\mathfrak{E}^1$ (line) spanned by a_1, a_2 is given by $x = t_1 a_1 + t_2 a_2$, $t_1 + t_2 = 1$. We recognize here the well known parametric representation of the line through the two points.

If we consider $t = (t_o, \cdots, t_p)$ as a point in an Euclidean space $\mathfrak{E}'^{p+1}$ the set $\mathfrak{E}'^p = \{t \,|\Sigma t_i = 1\}$ is topologized and we may state:

(12.2) *The correspondence $\theta : x \leftrightarrow t$ is a homeomorphism between $\mathfrak{E}^p$ and $\mathfrak{E}'^p$.*

In fact θ is independent of the coordinate system in $\mathfrak{E}^n$. Let it be so chosen that $\mathfrak{E}^p$ is the space $x_{p+1} = \cdots = x_n = 0$. This reduces property (12.2) to the same for $p = n$. Now for $p = n$ (12.2) assumes the explicit form

$$x_j = \Sigma t_i a_{ij},\ \Sigma t_i = 1\ (i = 0, \cdots, n; j = 1,2, \cdots, n),$$

with the determinant of the coefficients of the t_i different from zero. This system may be solved linearly for the t_i in terms of the x_j and the solution is unique. From this the proposition follows immediately.

13. The *geometric p-simplex*, written σ_g^p, is defined as follows: σ_g^o is a point; σ_g^p, $p > 0$, is defined in terms of $p + 1$ independent points $a_o, \cdots, a_p$ of $\mathfrak{E}^p$ as the set of all points x such that

(13.1) $$x = \Sigma t_i a_i,$$

(13.2) $$0 < t_i < 1, \Sigma t_i = 1.$$

The set σ_g^1 is an open interval and σ_g^2 is a triangle (perimeter excluded).

If we replace in (13.2) for $i = i_1, \cdots, i_{p-q}$ the inequality $0 < t_i$ by $0 = t_i$ there is obtained the σ_g^q spanned by the a_{i_h}, $i_h \neq i_1, \cdots, i_{p-q}$. This σ_g^q is called a *q-face* of $\bar{\sigma}_g^p$. The a_i are the zero-faces and are known as *vertices*. The one-faces are often called *edges*.

The *closed simplex* is the point-set closure $\bar{\sigma}_g^p$ and it is the union of the points of σ_g^p and of all its faces.

The dimensions, orientations and the like are defined as for σ^p with the vertices $a_0, \cdots, a_p$, and the various designations remain the same.

(13.2) *σ_g^p is a p-cell and $\bar{\sigma}_g^p$ is a closed p-cell and therefore a compactum.*

For the second is a bounded convex subset of an $\mathfrak{E}^p$ (I,10.3).

It is of interest to correlate the orientation of σ_g^p with the orientation of the space $\mathfrak{E}^p$ of σ_g^p by means of coordinate systems. If $\{x_i\}$, $\{y_i\}$ are any two such systems (not necessarily orthogonal) they are related by a transformation

(13.4) $$y_i = \Sigma \alpha_{ij} x_j + \beta_i, \qquad D = |\alpha_{ij}| \neq 0.$$

The transformation is *direct* if $D > 0$, *inverse* if $D < 0$. The orientation process is now the same as for σ^p with direct [inverse] transformations in place of even [odd] permutations. An even [odd] permutation of the coordinates is a direct [inverse] transformation. Hence the order of the coordinates defines an orientation of the space.

Let now a_0 be the origin of coordinates and a_i the point at unit distance from a_0 on the x_i-axis. Let $\sigma_g^p = a_0 \sigma_g^{p-1}$, where σ_g^{p-1} is the face opposite a_0. The orientation of σ_g^p defines the one of σ_g^{p-1}, hence the order in which the coordinates are named. Conversely given $\sigma_g^p = a_0 \cdots a_p \subset \mathfrak{E}^p$ there is defined in unique manner a coordinate system $\{x_i\}$ related to it in the preceding way. If $\sigma_g'^p$ is a second geometric simplex in $\mathfrak{E}^p$ and $\{y_i\}$ the associated coordinate system then σ_g^p, $\sigma_g'^p$ are said to be *concordantly* or *nonconcordantly oriented* accordingly as the related transformation (13.4) is or is not direct. The correlation between the orientation of $\mathfrak{E}^p$ and the orientation of its geometric p-simplexes σ_g^p is now complete. A σ_g^p which serves to orient $\mathfrak{E}^p$ by means of its coordinate system is sometimes called an *indicatrix* of the space.

14. Geometric complex. Consider now a simplicial complex $K = \{\sigma\}$ whose vertices $\{a_i\}$ are points of $\mathfrak{E}^p$. Suppose that the following conditions hold:

(a) The vertices of any simplex σ are independent, and hence they determine a $\sigma_g \subset \mathfrak{E}^r$.

(b) If σ, σ' are two distinct simplexes of K then the corresponding geometric simplexes σ_g, σ_g' are disjoint.

The resulting collection $K_g = \{\sigma_g\}$ is known as a *geometric complex*. The dimensions and all incidence relationships are carried over from K to K_g in the obvious way. The union of all the σ_g of K_g is known as a *simplicial polyhedron* or merely *polyhedron* written $|K_g|$. More generally the same designation $|K_g|$ will be applied even when K_g is merely a set of geometric simplexes and not necessarily a geometric complex. Thus $|St\ \sigma_g|$ denotes the set of all points in the simplexes of the star of σ_g.

If K' is isomorphic with K we say also that K' is isomorphic with K_g. Similarly, if K_g' and K_g'' are both isomorphic with K we define them as *isomorphic* with one another. The relationship is in every way the same as in (3).

Examples. The decomposition of the surface of an octahedron into triangles, edges, vertices gives rise to a K_g^2. A cube partitioned into elements by its diagonal planes is turned into a K_g^3.

It is often convenient to introduce barycentric coordinates in $|K_g|$ in the following way. For each vertex a_h we choose a real variable t_h. If x is any point of $\sigma_g = a_i \cdots a_j \in K_g$, the values of $t_i, \cdots, t_j$ at x are those of the corresponding barycentric coordinates as to $a_i, \cdots, a_j$ in the space of σ_g, and the other t_h are chosen equal to zero at the point. Thus every point x of $|K_g|$ is represented by relations

$$(14.1) \qquad x = \Sigma t_h a_h, \qquad \Sigma t_h = 1,\ t_h \geq 0.$$

This correspondence is readily seen to be one-one between $|K_g|$ and the resulting point set T of the space of the variables $(t_1, \cdots)$. In view of (13.3) it is also topological on each $\bar{\sigma}_g$ and hence throughout $|K_g|$.

(14.2) *Every simplicial complex K is isomorphic with some geometric complex K_g.*

Let $\{a_i\}$ be the vertices of K and let r be their number. In an $\mathfrak{E}^r$ let A_i be the point whose coordinates are the Kronecker deltas δ_{ij} ($\delta_{hk} = 0$ for $h \neq k$, $\delta_{hh} = 1$). Corresponding to any $\sigma = a_i \cdots a_j \in K$ construct the geometric simplex $\sigma_g = A_i \cdots A_j$ in $\mathfrak{E}^r$. The set of all these σ_g gives rise to a K_g isomorphic with K. Indeed (a) holds since the matrix of the coordinates of any σ_g^p contains a determinant of order $p + 1$ whose value is unity. As for (b) if $\sigma_g^p \neq \sigma_g^q$ with $p \geq q$ then one of the coordinates vanishes in σ_g^q but not in σ_g^p and so they are disjoint. Thus

K_g is a geometric complex. Its isomorphism with K is a consequence of the construction.

(14.3) *If the complexes K_g and K'_g are isomorphic then their polyhedra $|K_g|$, $|K'_g|$ are homeomorphic.*

Let $\{a_i\}$, $\{b_i\}$ be the vertices of K_g, K'_g (a_i associated with b_i under the isomorphism). To $x \in |K_g|$ with the barycentric coordinates t_i relative to the vertices a_i there corresponds $x' \in |K'_g|$ with the same coordinates relative to the b_i and $x \to x'$ manifestly defines a homeomorphism between $|K_g|$ and $|K'_g|$.

(14.4) *If K_g is finite then the polyhedron $|K_g|$ is a compactum.*

For it is the union of the compacta $\bar{\sigma}_i$ whose number is finite.

(14.5) *$|Cl\, \sigma_g|$ is closed in $|K_g|$ and $|St\, \sigma_g|$ is open in $|K_g|$.*

Since $Cl\, \sigma_g$ is a geometric complex $|Cl\, \sigma_g|$ is compact, hence closed in $|K_g|$. If σ'_g is not in $St\, \sigma_g$ then σ'_g does not meet $|St\, \sigma_g|$. Since the union of all such σ'_g is compact, it is closed in $|K_g|$, and so its complement $|St\, \sigma_g|$ is open.

15. Topological simplicial complexes. Let $\mathfrak{K} = \{\zeta_i^p\}$ be a collection of sets with the following properties:

(a) the ζ_i^p are disjoint subsets of a metric space;

(b) ζ_i^p is a p-cell;

(c) $\bar{\zeta}_i^p - \zeta_i^p = S_i^{p-1}$ is a $(p-1)$-sphere and is the union of sets ζ_j^q, $(q = 0, \cdots, p-1)$, called *faces* of ζ_i^p in the same number for each q as σ^p has q-faces; S_i^{-1} is the null set;

(d) the spheres S_i^{p-1}, $p \geq 1$, are all distinct.

The collection $\mathfrak{K}$ is called a *topological simplicial complex.* Its dimension is sup dim ζ, and the union of the sets ζ's is written $|\mathfrak{K}|$. The justification for the way of naming $\mathfrak{K}$ is found in:

(15.1) *There exists a geometric complex $K_g = \{\sigma_{gi}^p\}$ such that:* (a) *$\sigma_{gi}^p \leftrightarrow \zeta_{gi}^p$ is one-one;* (b) *there exists a homeomorphism $\varphi: |K_g| \to |\mathfrak{K}|$ such that $\varphi\sigma_i^p = \zeta_i^p$.*

Let $\mathfrak{K}^p$ denote the p-section of $\mathfrak{K}$ or sets of ζ_i^q, $q \leq p$, and let $(15.1)_p$ denote (15.1) for $\mathfrak{K}^p$. Then $(15.1)_o$ is obvious, and so we assume $(15.1)_{p-1}$ and prove $(15.1)_p$. Thus we have a K_g^{p-1} say in $\mathfrak{E}^r$, and φ defined in $|K_g^{p-1}|$ such that $\varphi\zeta_i^q = \sigma_i^q$ for $q \leq p$. Moreover referring to the proof of (14.2) we may suppose that the vertex A_i of K_g^{p-1} is the image of ζ_i^o and is the point with the coordinates δ_{ij}. Now if $\zeta_{i_o}^o, \cdots, \zeta_{i_p}^o$ are the zero-faces of ζ_i^p there exists in $\mathfrak{E}^r$ a simplex $\sigma_{gi}^p = A_{i_o} \cdots A_{i_p}$, and it is a consequence of $(15.1)_{p-1}$ that $\varphi S_i^{p-1} = |\mathfrak{B}\, \sigma_{gi}^p|$. The extension of φ to a homeomorphism over ζ_i^p to σ_i^{pg} is elementary and $(15.1)_p$, hence (15.1), follows.

Notice that under our definition no two distinct elements of $\mathfrak{K}$ have the same vertices (the ζ_i^0). This rules out for instance the partition of a circumference by two points, or of a sphere by a spherical triangle. A little care is thus necessary in triangulating. It is clear that the triangulations of §1 may be chosen such as to make up topological simplicial complexes.

Consider the triangulation of a sphere S^2 caused by three mutually perpendicular planes. As we vary the planes in all possible ways we obtain a collection of topological two-complexes covering S^2. These complexes $\mathfrak{K}^2$ determine the same set $|\mathfrak{K}^2| = S^2$ and it will be one of our major tasks to investigate their common topological properties, i.e. the properties of $\mathfrak{K}^2$ which are unchanged when $\mathfrak{K}^2$ varies while $|\mathfrak{K}^2|$ remains fixed. The chief properties of this nature are associated with the concept of homology.

Returning to the couple $\mathfrak{K}$, K_g one may introduce in the obvious way the concept of isomorphism of $\mathfrak{K}$ with K_g, hence of $\mathfrak{K}$ with a simplicial isomorph K of K_g. In all the questions dealing with $\mathfrak{K}$, K_g, K as distinguished from the sets $|\mathfrak{K}|$, $|K_g|$ and their topology, the three complexes are interchangeable. For this reason we shall refer henceforth loosely to any one of the three types as a "simplicial complex." Similarly, even as regards topological questions, $|\mathfrak{K}|$ and $|K_g|$, hence also $\mathfrak{K}$ and K_g, are interchangeable. For this reason we refer henceforth loosely to $\mathfrak{K}$ or K_g as "a geometric complex" and to $|\mathfrak{K}|$ or $|K_g|$ as "a polyhedron." Finally the subscripts g will be omitted throughout and we shall write K,σ instead of K_g, σ_g.

§5. Calculation of the Betti Numbers. The Euler-Poincaré Characteristic.

16. Let K be a finite simplicial complex and let us calculate its Betti numbers mod π. For convenience we omit the subscript π throughout the calculation.

We will set

$$[\sigma_i^{p+1}:\sigma_j^p] = \eta_{ij}(p), \qquad \eta(p) = \|\eta_{ij}(p)\|$$

and denote by α^p the number of σ_i^p and by ρ^p the rank of $\eta(p)$ mod π. The matrix $\eta(\rho)$ is known as the *p-th incidence matrix* of the complex.

We recall the basic inclusion for the groups $\mathfrak{C}$, $\mathfrak{Z}$, $\mathfrak{F}$:

$$\mathfrak{C}^p \supset \mathfrak{Z}^p \supset \mathfrak{F}^p. \tag{16.1}$$

Take in $\mathfrak{C}^p$ a maximal set $\{\delta_i^p\}$ of chains independent mod $\mathfrak{Z}^p$ (no linear combination of the δ_i^p is a cycle). Let $\mathfrak{D}^p$ be the subspace of $\mathfrak{C}^p$

spanned by the δ_i^p. It is a consequence of the definition of $\mathfrak{D}^p$ that $\mathfrak{C}^p$ admits the direct sum decomposition

$$(16.2) \qquad \mathfrak{C}^p = \mathfrak{D}^p + \mathfrak{Z}^p, \qquad \mathfrak{D}^p \cap \mathfrak{Z}^p = 0$$

and therefore

$$(16.3) \qquad \alpha^p = \dim \mathfrak{C}^p = \dim \mathfrak{D}^p + \dim \mathfrak{Z}^p.$$

We may also interpret (16.2) as meaning

$(16.4)_p$ *Every C^p is a linear combination of the δ_i^p plus a cycle.*

Furthermore

$(16.5)_p$ *No linear combination of the δ_i^p is a cycle.*

If we set $\beta_i^p = F\delta_i^{p+1}$ then it is a consequence of $(16.4)_{p+1}$ that every FC^{p+1} is a linear combination of the β_i^p, and a consequence of $(16.5)_{p+1}$ that the β_i^p are linearly independent. Hence the independent β^p span $\mathfrak{F}^p$, and therefore

(16.6) *The set $\{\beta_i^p\}$ is a base for $\mathfrak{F}^p$.*

Take now in $\mathfrak{Z}^p$ a maximal set $\{\gamma_i^p\}$, $i = 1,2, \cdots, R^p$, of elements linearly independent mod $\mathfrak{F}^p$. By definition they span a subspace $\mathfrak{S} \cong \mathfrak{H}^p$, and so their number is in fact $\dim \mathfrak{H}^p = R^p$. We now have the direct sum decomposition

$$\mathfrak{Z}^p = \mathfrak{S} + \mathfrak{F}^p,$$

and therefore

$$(16.7) \qquad \dim \mathfrak{Z}^p = \dim \mathfrak{H}^p + \dim \mathfrak{F}^p = R^p + \dim \mathfrak{F}^p.$$

Notice, for further reference, that corresponding to any π one may select simultaneous bases $\{\beta_i^p, \gamma_i^p, \delta_i^p\}$ for the chains mod π such that

$$F\delta_i^{p+1} = \beta_i^p, \qquad F\gamma_i^p = 0, \qquad F\delta_i^p = \beta_i^{p-1}.$$

Moreover $\{\gamma_i^p\}$ is a base for all the p-cycles mod π relative to homology, every $\beta_i^p \sim 0$ and no linear combination of the δ_i^p is a cycle mod π. In particular the number of cycles γ_i^p is R^p.

Returning to (16.7) we deduce from it

$$R^p = \alpha^p - \dim \mathfrak{D}^p - \dim \mathfrak{F}^p.$$

Since the number of δ_i^{p+1} and that of β_j^p are the same, $\dim \mathfrak{D}^{p+1} = \dim \mathfrak{F}^p$ and so

$$R^p = \alpha^p - \dim \mathfrak{F}^p - \dim \mathfrak{F}^{p-1}.$$

The space $\mathfrak{F}^p$ is spanned by the chains $F\sigma_i^{p+1} = \Sigma\eta_{ij}(p)\sigma_j^p$. The number of linearly independent chains of this type is the rank ρ^p of the matrix η^p. Therefore dim $\mathfrak{F}^p = \rho^p$ and finally, restoring the subscript π:

$$R_\pi^p = \alpha^p - \rho^{p-1} - \rho^p \tag{16.8$_p$}$$

which is the required expression for R_π^p.

Since there are no matrices $\eta(-1)$, $\eta(n)$ at the extremes $p = 0$, n we must set $\rho^{-1} = \rho^n = 0$.

If we multiply $(16.8)_p$ by $(-1)^p$ and sum for p we obtain the classical *Euler-Poincaré relation*

$$\Sigma(-1)^p\alpha^p = \Sigma(-1)^p R_\pi^p. \tag{16.9}$$

The common value of the two sides is called the *Euler-Poincaré characteristic*, or merely *characteristic* of K, written $\chi(K)$. We note explicitly the important property:

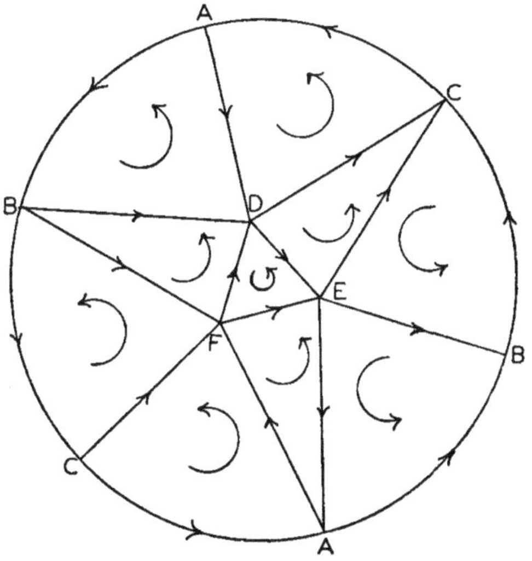

Figure 49

(16.10) *The expression* $\Sigma(-1)^p R_\pi^p$ *is independent of the modulus* π *and is equal to the characteristic* $\chi(K)$.

Historical remark. The original result of Euler was as follows: If Π is a convex polyhedron (in ordinary space) and V, E, F the number of its vertices, edges and faces then

$$\chi(\Pi) = V - E + F = 2.$$

If each face is triangulated by adding just one more vertex on the face and joining to its other vertices, there results a geometric complex

K_g for which one readily verifies that $\chi(K_g) = \chi(\Pi)$. We shall see in time that

$$R^0(K_g) = R^2(K_g) = 1, \qquad R^1(K_g) = 0$$

and so Euler's result is a consequence of (16.9).

The proof of the text is Poincaré's initial proof of the general formula (for $\pi = 0$).

17. Betti numbers of the projective plane. From the point of view of topology the projective plane is essentially a closed circle with the two end points of any diameter identified (II, 25.3). A triangulation of the configuration is given in Fig. 49 in which identified points have received the same designation. This triangulation is known to be one with the least number of triangles such that no two have the same vertices. We propose to calculate the Betti numbers of the topological complex K defined by Fig. 49.

The observation at the beginning of (11) gives at once $R^0 = R^0_\pi = 1$. Then from Fig. 49:

$$\alpha^0 = 6, \qquad \alpha^1 = 15, \qquad \alpha^2 = 10, \qquad \chi(K) = 1.$$

Let the orientations be as in Fig. 49 and let the triangles be denoted by $t_1, \cdots, t_{10}$. Any cycle γ^2 is an expression $\Sigma m_i t_i$. Now if t_i, t_j are the triangles adjacent to an interior edge such as DE, this edge has the coefficient $\pm (m_i - m_j)$ in $F\gamma^2$. Hence we must have $m_i = m_j$. Hence if $t_1 = DEF$, the coefficients of the adjacent triangles are all equal to m_1, and since the triangles left are adjacent to one of type DEF along an interior edge, all the coefficients m_i are equal. Thus $\gamma^2 = m\delta^2$, $\delta^2 = \Sigma t_i$.

On the other hand the figure yields at once

$$F\delta^2 = 2\gamma^1, \qquad \gamma^1 = AB + BC + CA, \qquad F\gamma^1 = 0.$$

Thus a two-cycle $\gamma^2 \neq 0$ may exist only in the system mod 2, and it is then necessarily δ^2. Therefore

$$R^2_\pi = 0 \text{ for } \pi \neq 2, \qquad R^2_2 = 1.$$

Hence finally

$$R^1_\pi = \chi(K) - R^0 - R^2_\pi = 0 \text{ for } \pi \neq 2, \qquad R^1_2 = 1.$$

18. Betti numbers of closed surfaces. Let Φ be a closed surface which for the present need not be orientable. The following results have already been established:

(18.1) $$R^0_2 = R^2_2 = 1.$$

(18.2) *The Betti numbers mod 2 and hence the characteristic* χ *are topologically invariant.*

(18.3) *For the orientable surface of genus p:*

$$\chi(\Phi) = 2 - 2p, \qquad R_2^1 = 2p.$$

We have indicated in (II, 20) that a certain orientation pattern, symbolized in Fig. 37 may be adopted for orientable, but not for non-orientable surfaces. Using this orientation pattern one may then paraphrase the reasoning of (II, 11.5) and show that

(18.4) *For an orientable surface* $R_\pi^2 = 1$, *whatever* π. *For a nonorientable surface* $R_\pi^2 = 0$ *when* $\pi \neq 2$, *and* $R_2^2 = 1$.

From this we deduce the following values for the Betti numbers:

Orientable surface of genus p:

$$(18.5) \qquad R_\pi^o = R_\pi^2 = 1, \qquad R_\pi^1 = 2p, \qquad \chi = 2 - 2p.$$

Nonorientable surface of characteristic χ:

$$(18.6) \quad R_2^o = R_2^2 = 1, \qquad R_2^1 = 2 - \chi;$$

$$R_\pi^o = 1, \qquad R_\pi^2 = 0, \qquad R_\pi^1 = 1 - \chi, \qquad \pi \neq 2.$$

§6. Relation Between Connectedness and Homology.

19. Roughly speaking throughout topology connectedness may be characterized by the Betti numbers R^o. This is the reason for the terms "connectivities" or "indices of connection" applied until recently to all the Betti numbers.

Consider first a complex K which is the union of r disjoint complexes $K_1, \cdots, K_r$. If C^p, γ^p are a chain and a cycle of K, then

$$C^p = \Sigma C_i^p, \qquad C_i^p \subset K_i; \qquad \gamma^p = \Sigma \gamma_i^p, \qquad \gamma_i^p \subset K_i.$$

Since the K_i are disjoint, $C^p = 0$ or $\gamma^p = 0$ when and only when respectively every $C_i^p = 0$ or every $\gamma_i^p = 0$. In other words the above decompositions are unique. In particular $F\gamma^p = \Sigma F\gamma_i^p = 0$, and hence $F\gamma_i^p = 0$, i.e. every γ_i^p is a cycle. Finally if $\gamma^p = FC^{p+1}$, $C^{p+1} = \Sigma C_i^{p+1}$, then $(FC_i^{p+1} - \gamma_i^p) = 0$, hence $\gamma_i^p = FC_i^{p+1}$. Thus $\gamma^p \sim 0$ implies that every $\gamma_i^p \sim 0$ and the converse is obviously true.

Let us use the substitution $\gamma \to \Gamma$ to pass from the cycles to their homology classes. The preceding remarks make it clear that there is a one-one correspondence between $\{\Gamma^p\}$ and the sets $(\Gamma_1^p, \cdots, \Gamma_r^p)$ and that it associates zeros and sums as in group products. Hence the relation

$$(19.1) \qquad \mathfrak{H}^p(K, G) \cong \mathfrak{H}^p(K_1, G) \times \cdots \times \mathfrak{H}^p(K_r, G).$$

It implies the following relation for the Betti numbers:

$$R^p_\pi(K) = \Sigma R^p_\pi(K_i). \tag{19.2}$$

20. We appear to be far from connectedness, but we come to it presently. Starting this time with the assigned complex K, consider a maximal set $\{a_i\}$ of vertices of K such that:

(a) every vertex b of K may be joined to one of the vertices a_i by a sequence $\zeta = a_i, a_1', \cdots, a_s', b$ in which consecutive vertices are joined by a σ^1 in K;

(b) No two a_i, a_j may be joined by a sequence such as ζ.

Let K_i be the set of all the simplexes of K having a vertex b which may be joined to a_i as described in (a). If $\sigma \in K_i$ and $\sigma' < \sigma$ then $\sigma' \in K_i$. Hence K_i is a subcomplex of K. Moreover it is a consequence of their definition that the K_i are disjoint and that K is their union. The K_i are known as the *components* of K. If there is only one, K is said to be *connected.* It is then characterized by the property that each vertex may be joined to a fixed vertex, and hence to any other vertex, by a sequence ζ.

If K is geometric, the polyhedron $|K_i|$ is arc-wise connected, hence connected and the $|K_i|$ are disjoint. Hence the sets $|K_i|$ are the point-set components of the polyhedron $|K|$. In particular if K is connected as a complex, $|K|$ is connected as a point-set, and conversely. Hereafter we shall merely say "connected" and "components" without distinguishing between the two types.

Since the components K_i of K are disjoint and K is their union the relations (19.1) and (19.2) hold for them.

Referring to (11) we see that K_i is cyclic in the dimension zero. Therefore $R^o_\pi(K_i) = 1$, and $R^o_\pi(K) = r$, the number of components. Moreover the integral group $\mathfrak{H}^o(K_i)$ is a free group on one generator and hence $\mathfrak{H}^o(K)$ is a free group on r generators.

To sum up we may thus state the

(20.1) **Theorem.** *The integral group $\mathfrak{H}^o(K)$ is a free group on r generators, where r is the number of components of K, and likewise, when K is geometric, the number of point-set components of the polyhedron $|K|$. Moreover $r = R^o_\pi$. Thus the numbers R^o_π are all equal and their common value, written R^o, is a topological invariant of the polyhedron $|K|$ when K is geometric.*

The following two special properties are singled out for later reference:

(20.2) *When K is connected it is cyclic in the dimension zero.*

(20.3) *When K is connected a zero-cycle* $\gamma^o \sim 0$ *if and only if its index* $KI(\gamma^o) = 0$.

We have already proved (20.2). Regarding (20.3), we find from (a) that if a_1 is the vertex of $K_1 = K$ there considered, then $\gamma^o \sim KI(\gamma^o)a_1$ and $a_1 \not\sim 0$. Hence $\gamma^o \sim 0$ when and only when its index is zero.

Examples. The preceding properties are readily verified for closed surfaces, for $Cl\ \sigma^n$ and for $\mathfrak{B}\ \sigma^{n+1}$, $(n > 0)$, for the disks and for the projective plane.

Another simple application is the determination of the Betti numbers of a linear graph K. According to what precedes we may as well assume K connected. Then $R^o = 1$ and

$$\chi(K) = \alpha^o - \alpha^1 = 1 - R^1_\pi.$$

Hence $R^1_\pi = 1 - \chi(K)$ is independent of π and is again written R^1. If $R^1 = 0$, there are no one-cycles, hence no closed paths in the graph, which is then a *tree* (II, 16).

The definition of "algebraic" connectedness (i.e. in the same sense as at the beginning of the present number) is applicable only to complexes. We shall require a definition valid for an arbitrary collection $\Lambda = \{\sigma\}$ of simplexes. We say then that Λ is *connected* whenever given any pair $\sigma, \sigma' \in \Lambda$ there is a sequence $\zeta = \sigma, \sigma_1, \cdots, \sigma_s, \sigma'$ of elements of Λ in which consecutive elements are incident. Such a sequence is said to be *connecting* σ, σ' in Λ. It is readily shown that: (a) when Λ is a complex the definition of algebraic connectedness is equivalent to the previous one; (b) the algebraic connectedness of the collection Λ is equivalent to the connectedness of the point-set $|\Lambda|$, the point-set union of the simplexes of Λ.

§7. Circuits.

21. This type of complex offers an interesting example of the relation between the structural properties and homology.

Definition. A simple n-circuit, or merely an *n-circuit,* is a complex $K = \{\sigma\}$ with the following properties:

(a) every σ^{n-1} of K is a face of two σ^n's;

(b) the set $\Lambda = \{\sigma^{n-1}_i, \sigma^n_j\}$ is connected;

(c) no proper subcomplex of K possesses the preceding two properties. This may also be expressed as: K is irreducible with respect to these two properties.

Examples: The projective plane of (17), the triangulated sphere, the triangulated two-sided disk are two-circuits. It is an elementary matter to show that $\mathfrak{B}\ \sigma^n$, $n > 1$, is an $(n - 1)$-circuit.

We shall now assume that K is an n-circuit and prove a number of its properties.

(21.1) *Every σ is a face of a σ^n.*

(21.2) $\dim K = n$.

(21.3) *K is connected or equivalently $R^o(K) = 1$.*

Suppose (21.1) false and let σ^p be a simplex of the highest dimension for which (21.1) fails. Then σ^p is not a proper face of any simplex and hence $K - \sigma^p$ is a proper subcomplex of K satisfying (a) and (b). Since this contradicts (c), (21.1) is true. Properties (21.2) and (21.3) follow immediately from (21.1) and the definition of the circuit.

Further results will now be obtained by examining K for n-cycles. For this purpose a preliminary construction will be advantageous. We construct a sequence $K_1, K_2, \cdots, K_s$ of subcomplexes as follows: $K_1 = Cl\, \sigma_1^n$; K_{r+1} is the union of the closures of all the n-simplexes having an $(n-1)$-face in K_r. Every K_r satisfies (b) and in view of (a) the process terminates only when the last complex K_s satisfies (a). Therefore by (c) $K_s = K$.

Let us now complete the construction as follows: σ_1^n is oriented in a definite way. Assume all the n-simplexes of K_r oriented; then if $\sigma^n \in K_{r+1} - K_r$, it has some $(n-1)$-face in common with an n-simplex of K_r. One such σ^{n-1} is selected; it is incident with a single $\sigma'^n \in K_r$ and we orient σ^n so that σ^n, σ'^n are oppositely related to σ^{n-1}. As a consequence

$$(21.4) \qquad [\sigma^n : \sigma^{n-1}] = -[\sigma'^n : \sigma^{n-1}] = \pm 1.$$

The process just described is called *orientation by propagation.* Suppose now that

$$(21.5) \qquad C^n = \Sigma g_i \sigma_i^n$$

is a cycle over G. In the above notations if g, g' are the coefficients of σ^n, σ'^n in C^n then FC^n will contain the term $\pm(g - g')$. Thus we must have $g = g'$. Therefore if the coefficients of the σ_i^n in K_r are equal this must also hold in K_{r+1}. Since this is trivial in K_1, it holds in every K_r and therefore in K. Hence if

$$\gamma^n = \Sigma \sigma_i^n,$$

every n-cycle is of the form $g\gamma^n$. There are now two possibilities:

I. At the termination of the propagation process, i.e. in K, every σ^{n-1} is oppositely related to the two adjacent n-simplexes. Then K is said to be *orientable.* Here $F\gamma^n = 0$, or γ^n is a cycle, and so K is cyclic in the dimension n.

II. At the end of the propagation process some σ^{n-1}'s are *not* oppositely related to their adjacent n-simplexes. Then K is said to be *nonorientable.* If $\{\sigma_i'^{n-1}\}$ are those behaving in the manner just described, $F\gamma^n = 2\Sigma(\pm\sigma_i'^{n-1})$. Hence γ^n is a cycle mod 2 and K has no other n-cycles whatsoever.

The Betti numbers for the dimension n are:

$$K \text{ orientable: every } R_\pi^n = 1;$$

$$K \text{ nonorientable: } R_2^n = 1,\ R_\pi^n = 0 \text{ for } \pi \neq 2.$$

Thus these Betti numbers are sufficient to separate the two cases.

(21.6) *In the definition of the circuit the connectedness of the collection* Λ *may be replaced by the condition* $R_2^n = 1$.

Let $\{\Lambda_i\}$ be the components of Λ. Owing to (a) the sum γ_i^n of the n-simplexes of Λ_i is an n-cycle mod 2. Suppose that there could exist a relation

$$\Sigma\epsilon_i\gamma_i^n \sim 0, \tag{21.7}$$

where the ϵ_i are not all zero mod 2. Thus let $\epsilon_1 = 1$. Since $\dim K = n$, in (21.7) $\sim$ may be replaced by $=$. Since γ_i^n,γ_j^n, $i \neq j$, have no common simplexes their elements cannot cancel one another. Hence (21.7) implies $\gamma_1^n = 0$. As this does not hold all the ϵ_i must be zero. Hence $R_2^n \geq r$, where r is the number of components Λ_i. Now we have already shown that when $r = 1$ (Λ is connected) $R_2^n = 1$. Conversely $R_2^n = 1$ implies $r = 1$ or that Λ is connected. This proves (21.6).

We have seen that when the n-circuit K is orientable, it possesses an integral n-cycle γ^n such that every n-cycle of K is of the form $g\gamma^n$. We shall refer to a cycle such as γ^n as a *fundamental cycle* for the circuit. Suppose that δ^n is a second fundamental cycle $\neq 0$. Then $\delta^n = p\gamma^n$, $\gamma^n = q\delta^n$ where p,q are integers. As a consequence, $(1 - pq)\delta^n = 0$, and as $\delta^n \neq 0$, we must have $pq = 1$, $p = q = \pm 1$. Thus δ^n can only be $\pm\gamma^n$. It is clear also that if γ^n is fundamental so is $-\gamma^n$. In other words, there are exactly two fundamental cycles $+\gamma^n$ and $-\gamma^n$. To orient the circuit is to assign to it one of γ^n, $-\gamma^n$. This may be done by specifying the orientation of any $\sigma^n \in K$. This σ^n is then called an *indicatrix* of the circuit. The parallel with the orientation of Euclidean spaces in (13) is obvious.

Among the circuits of the examples the only one which is nonorientable is the projective plane.

PROBLEMS

General remark. All the configurations considered in the problems below are supposed to be decomposable into the elements of a suitable topological complex K and the homology characters mentioned are those of K. A further assumption is made that these characters are independent of K. This is an actual theorem which will be proved later (V, 8.4). In each case the reader is therefore free to select a complex K as convenient as possible.

1. Devise actual triangulations for the plane disk Δ_p and the surface Φ_p of genus p and by means of the triangulations calculate the Betti numbers of Δ_p and Φ_p.

2. Let P be a right circular cylinder with the circular bases B, B' oriented alike. Let B and B' be identified with orientations preserved. The result is a nonorientable surface Ψ known as a *Klein bottle*. Calculate its Betti numbers.

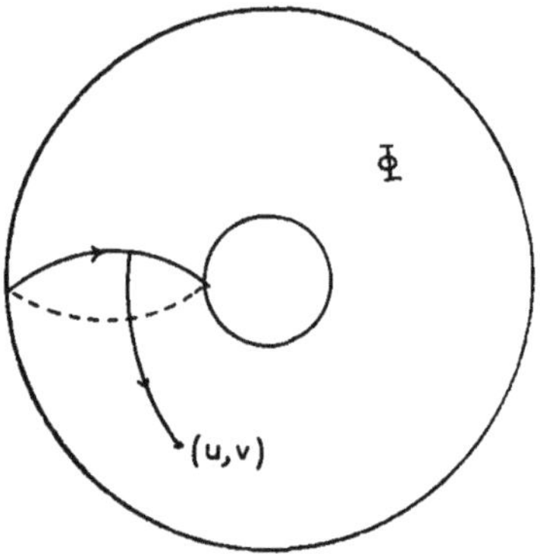

Figure 50

3. Match the opposite faces of a cube in all possible ways and calculate the Betti numbers of the solids thus obtained.

4. In a triangulated torus one of the triangles is replaced by the complement of the triangle DEF of the triangulation of the projective plane in Fig. 49. Find the homology characters of the new surface.

5. In an ordinary sphere S^2 two distinct points A, B are identified. Find the Betti numbers of the new configuration. Generalize in two ways. (a) $2n$ distinct points $A_1, B_1, \cdots, A_n, B_n$ are chosen and A_i, B_i are identified; (b) m distinct points are identified.

6. The surface Φ of genus p of (I, 24), is supposed to be in ordinary space $\mathfrak{E}^3$ and limits there a region Ω. The set $\bar{\Omega} = \Omega \cup \Phi$ is decomposed into the elements of a simplicial complex K^3. Find the Betti numbers of K^3.

7. Define projective three-space P^3 after the manner of (17) and find its Betti numbers.

8. Let Φ, Φ' be two *ring surfaces* (surfaces of genus 1) in $\mathfrak{S}^3$ and let T, T' be the solids which they bound. Let T, T' be referred to angular parameters u,v and u', v' where u, u' denote the parallel angles and v, v' the meridian angles. (In Fig. 50 a sketch of Φ is shown.) The mapping $t : u' = u$, $v' = v + mu$ (m a fixed integer) is a topological trans-

Figure 51

formation $\Phi \to \Phi'$. Let the points corresponding under t be identified thus giving rise to a solid $\Sigma = T \cup T' \cup \Phi$. Find the Betti numbers of Σ.

9. Let the rectangle $ABCD$ be decomposed into k equal strips after the manner of Fig. 51 and let AB be identified with CD, to produce a Möbius strip, so that the points numbered alike on AB and CD coincide. The Möbius strip is severed along the lines $(1, k - 1)$, $(2, k - 2)$, etc. What is the effect of this operation?

10. Show that the two-simplexes of the four-sphere $\mathfrak{B}\sigma^5$ make up two projective planes triangulated as in Fig. 49, each containing every one-simplex of the sphere.

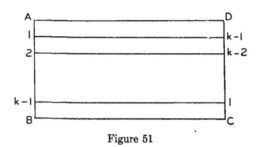

IV. Transformations of Complexes. Simplicial Approximations and Related Questions

In the present chapter we take up transformations of complexes both on the geometric and the algebraic side. Particular attention is paid to simplicial approximations of mappings. Many applications are made notably to Brouwer's fixed point theorem, to his degree, and finally to H. Hopf's classification of mappings of n-spheres on n-spheres.

§1. SET-TRANSFORMATIONS. CHAIN-MAPPINGS.

1. Let $K = \{\sigma\}$, $L = \{\zeta\}$ be two complexes. A natural type of transformation suggests itself immediately, namely an assignment to each $\sigma \in K$ of a set of simplexes $\zeta \in L$ which may even make up a subcomplex of L. A transformation of this nature is known as a *set-transformation* of K into L. It is a transformation of K into L, in the ordinary sense, when they are both merely viewed as sets of elements. Such transformations are of rather minor interest for us. In the more interesting cases t will have the following property: if $\sigma' < \sigma$, then $t\sigma' \subset Cl\ t\sigma$. We shall then say that t is *closed*.

Of far more importance are the transformations of the chains. Such an operation τ to be significant should consist first of all of a set of simultaneous homomorphisms of the groups of p-chains $\mathfrak{C}^p(K;G)$ of K into the corresponding groups $\mathfrak{C}^p(L;G)$ of L. On the other hand, there is ample reason to believe that the properties of the chain groups which are topologically significant are all tied up with the homology groups. Hence it would be well to require that τ send respectively a chain and its boundary into a chain and its boundary. This suggests commutation with the boundary operator F and justifies the

Definition. A *chain-mapping* τ of K into L is a set of simultaneous homomorphisms of the integral groups $\mathfrak{C}^p(K)$ each into the integral group $\mathfrak{C}^p(L)$ of equal dimension, which commutes with the operator F : $\tau F = F\tau$.

Since τ is a homomorphism of groups on finite bases, it is completely defined by its values on the base elements σ. This means that it is completely defined by a set of relations

$$\tau\sigma_i^p = \Sigma a_{ij}(p)\zeta_j^p \tag{1.1}$$

where the $a_{ij}(p)$ are integers. The value of τ on any integral chain is defined by linear extension:

$$\tau\Sigma g_i\sigma_i^p = \Sigma g_i a_{ij}(p)\zeta_j^p. \tag{1.2}$$

For convenience set

$$[\sigma_i^p : \sigma_j^{p-1}] = \alpha_{ij}(p-1), \qquad [\zeta_i^p : \zeta_j^{p-1}] = \beta_{ij}(p-1). \tag{1.3}$$

Then commutation with F leads to the relations $\tau F\sigma_i^p = F\tau\sigma_i^p$ or

$$\Sigma a_{ij}(p-1)a_{jk}(p-1) = \Sigma a_{ij}(p)\beta_{jk}(p-1) \tag{1.4}$$

for all pairs (i,k) and all acceptable values of p.

We may now define τ for chains over any group G by linear extension, i.e. again by (1.2) where $g_i \in G$, and it is readily seen that even extended in this manner τ still commutes with F. Notice however that:

(1.5) *A n.a.s.c. for τ as just extended to be an isomorphism of every chain group of K with the corresponding group of L, is that it possesses that property for the integral groups.*

Since τ commutes with F it sends a cycle into a cycle and a bounding cycle into a bounding cycle. For if C^p is a cycle of K, then $FC^p = 0$, hence $F\tau C^p = \tau(FC^p) = 0$, or τC^p is a cycle. Moreover if C^p bounds: $C^p = FC^{p+1}$, both chains being in K, then $\tau C^p = \tau FC^{p+1} = F(\tau C^{p+1})$, so that τC^p also bounds. As a consequence

$$\tau\mathfrak{Z}^p(K;G) \subset \mathfrak{Z}^p(L;G), \qquad \tau\mathfrak{F}^p(K;G) \subset \mathfrak{F}^p(L;G).$$

Thus τ sends homologous cycles into homologous cycles. Therefore

(1.6) *τ induces a simultaneous homomorphism of the homology groups of K into the corresponding groups of L.*

(1.7) *If τ_1 is a chain-mapping $K \to L$* and *τ_2 a chain-mapping $L \to M$ then $\tau_2\tau_1$ is a chain-mapping $K \to M$.*

We recall the observation of (III, 8) regarding the case when G is a field: the groups are vector spaces over G, the homomorphisms are linear transformations and the isomorphisms nonsingular linear transformations.

2. Simplicial set-transformations and chain-mappings. These are the most significant operations on complexes from the standpoint of topology. It is in fact out of simplicial chain-mappings that the general concept of chain-mapping arose.

Let $\{A_i\}$ be the vertices of K and $\{B_j\}$ those of L. Let t be an assignment to each vertex A_i of K of a vertex $B_j = tA_i$ of L such that if $\sigma = A_i \cdots A_h \in K$ then $\zeta = (tA_i) \cdots (tA_h) \in L$. If we define $t\sigma = \zeta$, t is extended to a set-transformation $K \to L$. Such a set-transformation is said to be *simplicial.*

One may now define a simultaneous transformation $\tau : \mathfrak{C}^p(K;G) \to \mathfrak{C}^p(L;G)$ by

$$\tau(A_i \cdots A_h) = (tA_i) \cdots (tA_h) \tag{2.1}$$

and its linear extension. Notice that if $t\sigma = \zeta$ and dim σ = dim ζ then $\tau\sigma = \zeta$ also, while if dim $\zeta <$ dim σ, the simplex at the right in (2.1) has repeated vertices and so, under our conventions, it is zero as a chain.

(2.2) *τ is a chain-mapping.*

The chain-mapping τ is said to be *simplicial* and the related simplicial set-transformation t is known as its *carrier.*

The proof of (2.2) is very simple. It is only necessary to show that $\tau F\sigma = F\tau\sigma$ for every $\sigma \in K$. Choose the labels so that $\sigma = A_0 \cdots A_p$ and $tA_i = B_i$, where the B_i need not be distinct. If they are distinct

$$\begin{aligned} \tau F\sigma &= \tau\Sigma(-1)^q A_0 \cdots A_{q-1} A_{q+1} \cdots A_p \\ &= \Sigma(-1)^q B_0 \cdots B_{q-1} B_{q+1} \cdots B_p \\ &= FB_0 \cdots B_p = F\tau\sigma. \end{aligned}$$

If the vertices B_i are not distinct choose the labels so that $B_0 = B_1$. Then the first two terms in the last sum cancel and the others vanish since each contains repeated vertices. Hence $\tau F\sigma = 0$, $F\tau\sigma = 0$. Thus $\tau F\sigma = F\tau\sigma$ whatever σ and (2.2) follows.

(2.3) *A simplicial chain-mapping does not change the Kronecker indices of zero chains.*

For if $tA_i = B_j$ then $KI(A_i) = KI(B_j) = KI(\tau A_i)$. Hence $KI(\tau C^0) = KI(C^0)$.

§2. DERIVATION.

3. The very simple construction of (II, 3) for indefinite subdivision of complexes of dimension at most two does not extend beyond. For general complexes we shall have recourse for the same purpose to another process, derivation, which is somewhat more formal but possesses many algebraic advantages.

Let $K = \{\sigma\}$ be a simplicial complex, not necessarily geometric. Associate with each σ one new vertex $\hat{\sigma}$. Then corresponding to any set $\{\sigma_i, \cdots, \sigma_j\}$ of simplexes of K such that $\sigma_i < \cdots < \sigma_j$, introduce the simplex $\zeta = \hat{\sigma}_i \cdots \hat{\sigma}_j$. We shall refer respectively to $\hat{\sigma}_i$ and $\hat{\sigma}_j$ as the *first* and the *last* vertex of ζ. Let $K' = \{\zeta\}$. Any face of ζ is of the form $\zeta_1 = \hat{\sigma}_h \cdots \hat{\sigma}_k$ where $\{h, \cdots, k\}$ is a subset of $\{i, \cdots, j\}$. If $h, \cdots, k$ are written in the same order as they are found in $\{i, \cdots, j\}$ then $\sigma_h < \cdots < \sigma_k$ and therefore $\zeta_1 \in K'$. This shows that K' is a simplicial complex. It is called the *first derived* or merely the *derived* of K. The same process may be applied to K', etc., and we thus have the second, $\cdots$, the n-th derived, $\cdots$ of K, written K'', $\cdots$, $K^{(n)}$, $\cdots$.

The set-transformation $D : K \to K'$ which consists in replacing σ by the set of all the ζ's ending with $\hat{\sigma}$ is called *derivation.*

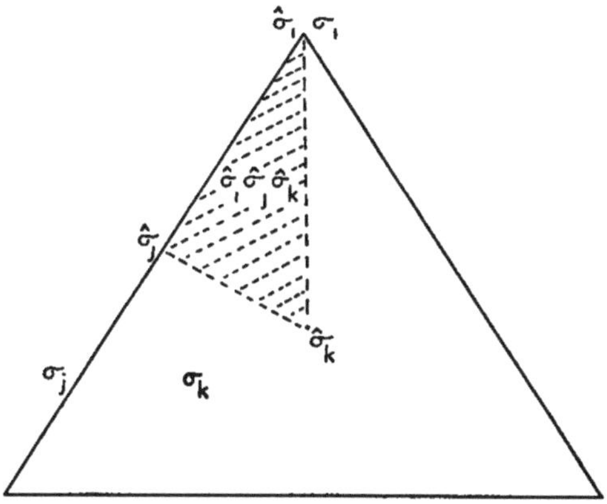

Figure 52

We note the following elementary properties:

(3.1) $$\dim K^{(n)} = \dim K.$$

(3.2) *D is a closed set-transformation and its defining relations are*

(3.3)$_o$ $$D\sigma^o = \hat{\sigma}^o = \sigma^o;$$

(3.3)$_p$ $$D\sigma^p = \hat{\sigma}^p \cup \hat{\sigma}^p D\mathfrak{B}\sigma^p, \qquad p > 0.$$

4. Barycentric subdivision. The operation D is particularly interesting when K is geometric. The vertex $\hat{\sigma}$ is then generally taken at the centroid of σ and K' is then called the *barycentric derived* or *barycentric subdivision* of K.

Generally speaking, the term *subdivision* as applied to a geometric complex $K = \{\sigma\}$ refers to a geometric complex $L = \{\zeta\}$ related to K as follows: (a) every ζ is in some σ; (b) every σ is the point-set union of the ζ's which it contains. This last property could also be replaced by: (b') the polyhedra $|K|$ and $|L|$ coincide. Thus K, L are different simplicial decompositions of the same polyhedron.

One will infer that derived complexes are subdivisions in the above sense. This is, in fact, one of several properties that we state together and prove afterwards.

(4.1) *A simplex ζ of K' ending in $\hat{\sigma}$ is contained in the simplex σ of K. Hence two simplexes of K' ending with distinct vertices are disjoint.*

(4.2) *If L is a subcomplex of K, the simplexes of $K^{(n)}$ contained in $|L|$ make up $L^{(n)}$.*

(4.3) *$K^{(n)}$ is a geometric complex.*

(4.4) *$K^{(n)}$ is a subdivision of K and any simplex σ of K is the union of the simplexes of $D^n\sigma$.*

Proofs of (4.1), $\cdots$, (4.4). It is clear that in the last three propositions it is sufficient to consider the case $n = 1$. Hence in all four we only need to consider K'. Let dim $K = p$ and let $(4.1)_p, \cdots$, denote $(4.1), \cdots$, for p. Since the propositions $(4.i)_0$ are trivial, we assume the $(4.i)_{p-1}$ and prove the $(4.i)_p$.

Regarding $(4.1)_p$ then we may assume that σ is a σ^p. Then $\zeta = \zeta'\hat{\sigma}^q\hat{\sigma}^p$, where $\sigma^q < \sigma^p$. Hence by $(4.1)_{p-1}$, $\zeta'\sigma^q$ is in σ^q. Therefore ζ is in the join of $\hat{\sigma}^p$ with σ^q, and hence it is in σ^p.

Under the hypothesis that $(4.2)_{p-1}$ holds, it is sufficient to prove $(4.2)_p$ for $L = Cl\,\sigma^p$. By (4.1) the only ζ's in $|\mathfrak{B}\sigma^p|$ are those ending in some $\hat{\sigma}^q$ where $\sigma^q < \sigma^p$. Thus we must show that the simplexes $\zeta\hat{\sigma}^p$, $\zeta \subset |\mathfrak{B}\sigma^p|$ fill up σ^p. Now if $x \in \sigma^p$, we may assume $x \neq \hat{\sigma}^p$. Then the ray from $\hat{\sigma}^p$ to x meets $|\mathfrak{B}\sigma^p|$ in a single point y. By hypothesis y is in some ζ of $(D\mathfrak{B}\sigma^p)'$, i.e. in some simplex of K' ending in a $\hat{\sigma}^q$, $\sigma^q < \sigma^p$. Hence the arc $\hat{\sigma}^p y$ is in σ^p and so is x. This implies (4.2).

We have seen that the simplexes of K' are disjoint. This suffices to show that K' is a geometric complex and (4.3) follows.

By (4.1) every simplex ζ of K' is in $|K|$, and all such ζ's fill up K. Hence K' is a subdivision of K. By (4.1) also σ is the union of the simplexes ζ ending in $\hat{\sigma}$, and these are precisely the ζ's making up $D\sigma$. Thus (4.4) is also proved.

4.5 *The mesh of the n-th barycentric derived $\to 0$ with increasing n.*

For the proof we require

4.6 *The diameter of a simplex is equal to the length of its longest edge.*

Let $\sigma = A_0 \cdots A_p \subset \mathfrak{E}^r$. If P is any point of $\mathfrak{E}^r$ and $\rho = \sup d(P,A_i)$ then the vertices A_i are all contained in the closed sphere $\mathfrak{S}(P,\rho)$. Since σ is convex it is wholly contained in the closed sphere and so no point of σ is farther than ρ from P. It follows that the longest segment in $\bar{\sigma}$ must have a vertex A_i as an end point, and the longest segment from A_i must be an edge.

Proof of (4.5). It is clear that we may assume $K = Cl\,\sigma^p$. It is to be recalled that the centroid G of σ^p is the point whose barycentric coordinates are all equal, their common value being then necessarily $\frac{1}{p+1}$. Let $\zeta = \hat{\sigma}^q\hat{\sigma}^r$, $\sigma^q < \sigma^r$, be any edge of K' and let λ be its length. Thus $\zeta \subset \sigma^r$. If σ^s is the face of σ^r opposite σ^q so that $\sigma^r = \sigma^q\sigma^s$, then a simple computation shows that the segment $\zeta_1 = \hat{\sigma}^q\hat{\sigma}^s$ contains ζ and so ζ_1 is the longest segment of $\bar{\sigma}^r$, hence of $\bar{\sigma}^p$, containing ζ.

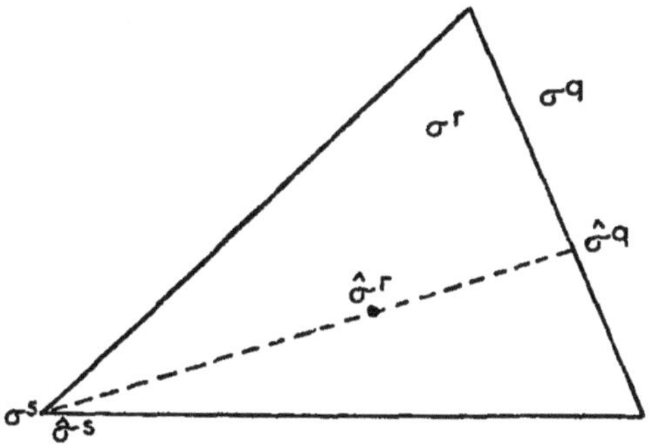

Figure 53

If μ is the length of ζ_1, then

$$\frac{\lambda}{\mu} \leq \frac{r-q}{r+1} \leq \frac{r}{r+1} \leq \frac{p}{p+1}.$$

Hence

$$\lambda \leq \frac{\mu p}{p+1} \leq \frac{p}{p+1} \times \text{mesh } K.$$

Therefore

$$\text{mesh } K' \leq \frac{p}{p+1} \times \text{mesh } K.$$

And so finally mesh $K^{(n)} \leq (\text{mesh } K) \times \left(\frac{p}{p+1}\right)^n$ which $\to 0$. This proves (4.5).

Thus barycentric subdivision provides a systematic method for subdividing a general complex into arbitrarily small pieces.

5. Side by side with the "geometric" operation of derivation D

there are two important companion chain-mappings: chain-derivation δ and an inverse τ of δ.

Chain-derivation. It is recursively defined as follows:

$$\delta\sigma^o = \hat{\sigma}^o = \hat{\sigma}^o; \qquad \delta\sigma^p = (-1)^p(\delta F\sigma^p)\hat{\sigma}^p = \hat{\sigma}^p\delta F\sigma^p, \qquad p > 0.$$

(The first expression for $\delta\sigma^p$ is the one in which the simplexes naturally appear written in the proper form of (3); the second however is often more convenient.) It is clear that $\delta\sigma^p$ is a chain of $D\sigma^p$. Let us show that δ is a chain-mapping. We merely have to prove

$$(5.1)_p \qquad F\delta\sigma^p = \delta F\sigma^p,$$

for every p. Now $(5.1)_o$ is obvious and so we use recursion on p. Assuming that $(5.1)_{p-1}$ holds, we find

$$F\delta\sigma^p = F\hat{\sigma}^p\delta F\sigma^p = \delta F\sigma^p - \hat{\sigma}^p F\delta F\sigma^p.$$

By the assumed property

$$F\delta(F\sigma^p) = \delta F(F\sigma^p) = \delta(FF\sigma^p) = 0$$

and $(5.1)_p$ follows. Hence δ is a chain-mapping.

The inverse τ of δ. Let t be an assignment to each vertex $\hat{\sigma}$ of K' of a vertex A of σ. If $\zeta = \hat{\sigma}_i \cdots \hat{\sigma}_j$, $\sigma_i < \cdots < \sigma_j$, then $t\hat{\sigma}_i, \cdots, t\hat{\sigma}_j$ are all vertices of σ_j. Therefore t determines a simplicial set-transformation $K' \to K$ and there is an induced chain-mapping τ. We shall prove the important relation

$$(5.2) \qquad \tau\delta = 1.$$

In view of this relation τ is a "left-inverse" of δ. In general it is not unique, except trivially when dim $K = 0$. For convenience we refer nevertheless to τ as an *inverse* of δ.

To prove (5.2) we must show that

$$(5.3)_p \qquad \tau\delta\sigma^p = \sigma^p$$

for every p. Since $(5.3)_o$ is obvious, we proceed by induction. Assume then $(5.3)_{p-1}$. Any simplex ζ^p of $\delta\sigma^p$ ends with $\hat{\sigma}^p$. Hence $\tau\zeta^p$ is $\pm\sigma^p$ or zero. Therefore $\tau\delta\sigma^p = \epsilon\sigma^p$, ϵ an integer. Applying F to both sides and recalling the hypothesis of the induction we have

$$F\tau\delta\sigma^p = \tau\delta F\sigma^p = F\sigma^p = \epsilon F\sigma^p.$$

Since $F\sigma^p \neq 0$, $\epsilon = 1$ and so $(5.3)_p$ holds. This proves (5.3).

Extension to higher derived. Let $K^{(s)} = D^sK$, the s-th derived of K. There is the obvious associated operation $\delta^s : K \to K^{(s)}$. Let $\{A_i\}$, $\{a_i\}$

be the vertices of K, $K^{(s)}$. By repetition of (4.1) we see that each simplex ζ of $K^{(s)}$ is contained in a simplex σ of K. Hence an assignment t to each a_i of a vertex $A_{j(i)}$ of K of the simplex σ of K containing a_i, defines a simplicial set-transformation $K^{(s)} \to K$ and there is an associated simplicial chain-mapping $\tau_s : K^{(s)} \to K$. This time again

$$\tau_s \delta^s = 1. \tag{5.4}$$

The proof is exactly the same as that of (5.3) and need not be repeated. Once more we refer to τ_s as an inverse of δ^s.

A noteworthy consequence of (5.4) is the following property which is a special case of a lemma due to Sperner:

(5.5) *Let $K = Cl\,\sigma^n$, $\sigma^n = A_0 \cdots A_n$. Let $K^{(s)}$ be any derived of K and let $\{a_i\}$ be its vertices. Consider any assignment $t : a_i \to A_{j(i)}$ such that $A_{j(i)}$ is a vertex of the face σ' of σ^n containing a_i. Then there is an odd number of simplexes $\zeta^n = a_{i_0} \cdots a_{i_n}$ of $K^{(s)}$ such that the ta_{i_h} are all distinct.*

The chain-mapping $K^{(s)} \to K$ induced by t is just an inverse τ_s of δ^s. If the lemma is incorrect and C^n is a chain mod 2 of $K^{(s)}$ we have $\tau_s C^n = 0$. In particular $\tau_s \delta^s(\sigma^n) = 0$ mod 2, whereas according to (5.4) $\tau_s \delta^s \sigma^n = \sigma^n \neq 0$.

Convention. Since no other subdivisions than the derived will ever be utilized in the sequel, we shall sometime say "subdivision" for "s-th derived," write then d, d^{-1} for δ^s, τ_s and refer to them as a chain-subdivision and an inverse of chain-subdivision. We notice explicitly the relation

$$d^{-1} \cdot d = 1 \tag{5.6}$$

which is merely (5.4) in the present notations.

§3. THE BROUWER FIXED POINT THEOREM.

6. (6.1) **Theorem.** *Every mapping of a closed n-cell (= n-disk) into itself has a fixed point* (Brouwer).

The proof merely requires the Sperner lemma in the form (5.5) and the existence of subdivisions of arbitrarily small mesh.

To illustrate the theorem, consider the unit parallelotope Π^n of $\mathfrak{E}^n$:

$$0 \leq x_i \leq 1, \qquad i = 1, 2, \cdots, n.$$

A mapping of Π^n into itself is given by a set of relations

$$x_i' = f_i(x_1, \cdots, x_n),$$

where the f_i are continuous and $0 \leq f_i \leq 1$ for all $(x_1, \cdots, x_n)$ in Π^n. A *fixed point* is any point satisfying the system

$$x_i = f_i(x_1, \cdots, x_n)$$

and Brouwer's theorem asserts that this system always possesses a solution.

For the segment an elementary but highly suggestive proof may be given in the following way. Let the segment be $0 \leq x \leq 1$ and let $f(x)$ be the transformation. Thus f is a continuous real function whose values on the segment are in the segment. As a consequence the graph $y = f(x)$, $0 \leq x \leq 1$, is an arc of curve γ in the square $0 \leq x, y \leq 1$ (Fig. 54) meeting every vertical segment in a single point. The fixed points

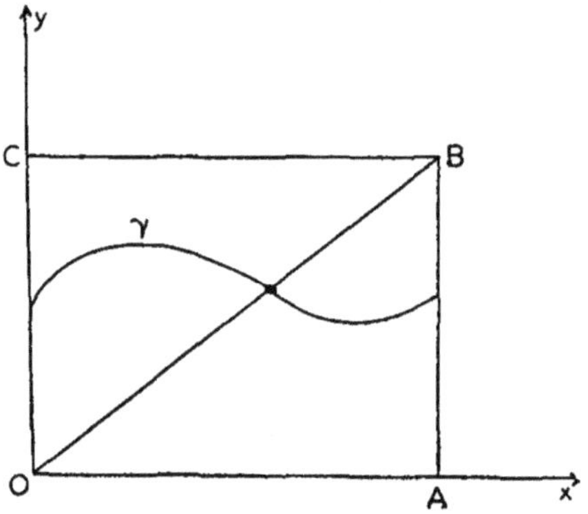

Figure 54

of f correspond to the intersections of γ with the diagonal OB and the proof that such points exist is elementary.

7. We pass now to the proof of Brouwer's theorem proper. Let $x_o, \cdots, x_n$ be barycentric coordinates for $\bar{\sigma}^n$, $\sigma^n = A_oA_1 \cdots A_n$. If $tx = x'$ the corresponding barycentric coordinates satisfy relations

$$x_i' = f_i(x_o, \cdots, x_n)$$

where the f_i are continuous. Let F_i be the subset of $\bar{\sigma}_n$ defined by

$$f_i(x_o, \cdots, x_n) \leq x_i. \qquad (7.1)$$

It is clear that the sets F_i are closed. Brouwer's theorem will follow if we can show that

(7.2) *The F_i intersect.*

For in that case there exists a pair x, $x' = tx$ such that $x_i' \leq x_i$ $(i = 1,2, \cdots, n)$. If in one of these relations we had $<$ and not $=$, we would have $1 = \Sigma x_i' < \Sigma x_i = 1$. Hence $x = x' = tx$, and x is a fixed point.

Let x be a point of the closed face $\bar{\sigma} = \overline{A_i \cdots A_j}$. In that face all the coordinates other than $x_i, \cdots, x_j$ are zero and $x_i + \cdots + x_j = 1$. Since $x_i' + \cdots + x_j' \leq 1$, we cannot have at the same time $x_i' > x_i, \cdots, x_j' > x_j$. Hence

$$\bar{\sigma} = \overline{A_i \cdots A_j} \subset F_i \cup \cdots \cup F_j.$$

Consider now an ϵ barycentric subdivision L of $K = Cl\,\sigma^n$ and let $\{B_h\}$ be its vertices. To each B_h there is assigned a vertex $A_{i(h)}$ by the following rule: if $B_h \subset \sigma$ above, there is a set F_i containing B_h and we assign to B_h the vertex $A_{i(h)} = A_i$. By Sperner's lemma there is a $\sigma_1^n = B_{h_0} \cdots B_{h_n}$ such that A_i is assigned to B_{h_i}. Thus $B_{h_i} \in F_i$. Hence whatever $\epsilon > 0$ there is a set of diameter $< \epsilon$ meeting every F_i. In particular if ϵ is less than the Lebesgue number of the covering $\{F_i\}$ of $\bar{\sigma}^n$, we conclude that the F_i intersect. This proves (7.2) and hence the theorem.

§4. Simplicial Approximation.

8. In the present section we shall discuss mappings of polyhedra into one another and prove a basic homotopy theorem regarding their approximation by barycentric mappings.

Let $K = \{\sigma\}$ with vertices $\{A_i\}$ and $L = \{\zeta\}$ with vertices $\{B_j\}$ be two geometric complexes and let $tA_i = B_{j(i)}$ be the defining relations of a simplicial set-transformation $K \to L$. Let also $\{s_i\}$ be barycentric coordinates for K. Thus every point x of the polyhedron $|K|$ is uniquely represented in the form

$$x = \Sigma s_i A_i, \qquad \Sigma s_i = 1, \qquad s_i \geq 0.$$

Consider now the operation t^* defined by

$$y = t^* x = \Sigma s_i (tA_i) = \Sigma s_i B_{j(i)}.$$

We propose to show that t^* is a (point-set) mapping $|K| \to |L|$.

If $x \in \sigma = A_i \cdots A_h$ then its coordinates s_k, $k = i, \cdots, h$ are $\neq 0$ and all the others vanish. Hence

$$y = s_i B_{j(i)} + \cdots + s_h B_{j(h)};\; s_i + \cdots + s_h = 1;\; s_i, \cdots, s_h > 0.$$

It follows that if $B_m', \cdots, B_n'$ are the distinct vertices in the set $B_{j(i)}, \cdots, B_{j(h)}$, then

$$y = s_m' B_m' + \cdots + s_n' B_n',$$

where s_r' is the sum of the coordinates corresponding to those vertices of the set $B_{j(i)}, \cdots, B_{j(h)}$, which coincide with B'_r. Suppose for instance (Fig. 55) that $\sigma = A_1A_2A_3$, $tA_1 = tA_2 = B'_1$, $tA_3 = B'_3$. Then $y = (s_1 + s_2) B'_1 + s_3B'_3$.

Returning to the general case we have

$$s'_m + \cdots + s'_n = s_i + \cdots + s_h = 1; \qquad s'_m, \cdots, s'_n > 0.$$

Hence $s'_m, \cdots, s'_n$ are the barycentric coordinates of a point of the simplex $\zeta = B'_m \cdots B'_n = t\sigma$. Thus if $x \in \sigma$ then $y \in t\sigma$. Hence t^* is a transformation $|K| \to |L|$. Its continuity is a consequence of the continuity of the barycentric coordinates (III, 14). Therefore t^* is a mapping. It is referred to as a *barycentric mapping* of K into L (thus underscoring its relationship to the two complexes), also as the *barycentric extension* of t.

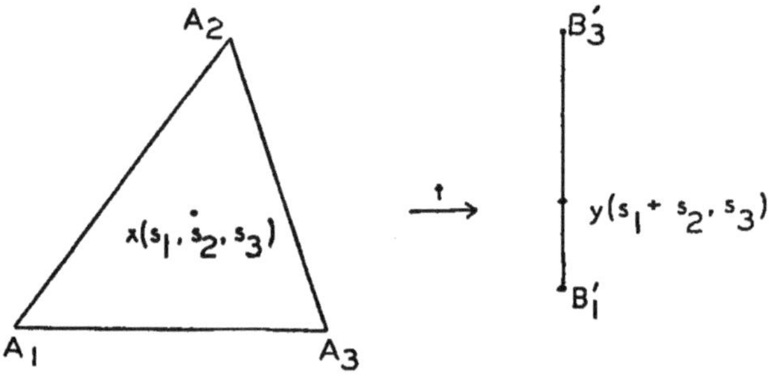

Figure 55

We have already shown that $t^*\sigma \subset \zeta$. It is readily seen that $s'_m, \cdots, s'_n$ may take all values between 0 and 1. Hence $\zeta \subset t^*\sigma$ and therefore $t^*\sigma = \zeta$. To simplify matters we shall write henceforth t for t^*, so that t denotes the barycentric mapping itself.

9. Let the notations remain as before, and consider the stars $St\ A_i$ of the vertices of K. If x is a point of the polyhedron $|K|$ then x is in some simplex σ of K. Hence if A_i is a vertex of σ, the point x is in the *star-set* $|St\ A_i|$ which is an open set of the polyhedron $|K|$. Therefore $\mathfrak{U} = \{|St\ A_i|\}$ is a finite open covering of $|K|$. Similarly $\mathfrak{B} = \{|St\ B_j|\}$ is a finite open covering of $|L|$.

(9.1) *If $A_i, \cdots, A_k$ are vertices of K then the stars $St\ A_i, \cdots, St\ A_k$ intersect when and only when $\sigma = A_i \cdots A_k$ is a simplex of K and their intersection is $St\ \sigma$.*

(9.2) *Similarly for the star-sets $|St\ A_i|, \cdots, |St\ A_k|$, the intersection being $|St\ \sigma|$.*

Proof of (9.1). Suppose that the stars intersect. Their intersection is a set of simplexes of K each of which has $A_i, \cdots, A_k$ as vertices. Hence each has $\sigma = A_i \cdots A_k$ as a face and so $\sigma \in K$. Conversely if $\sigma \in K$ then σ is in each of the stars and so they intersect.

Suppose now $\sigma \in K$. Any simplex of $St\,\sigma$ is of the form $\sigma\sigma'$ and it has $A_i, \cdots, A_k$ as vertices. Hence $\sigma\sigma'$ is in each of the stars $St\,A_i, \cdots, St\,A_k$. Therefore it is contained in their intersection and hence so is $St\,\sigma$. Conversely if σ'' is in the intersection it is in each of the intersecting stars and must have $A_i, \cdots, A_k$ as vertices. Hence $\sigma < \sigma''$, $\sigma'' \in St\,\sigma$ and so $St\,\sigma$ contains the intersection. This proves (9.1).

Proof of (9.2). If the sets $|St\,A_i|, \cdots, |St\,A_k|$ intersect and x is in the intersection, the simplex σ' of K containing x is in each $St\,A_i, \cdots, St\,A_k$. Thus the latter intersect and their intersection is the $St\,\sigma$ considered before. Moreover $\sigma' \in St\,\sigma$. Hence $x \in |St\,\sigma|$ and so $|St\,\sigma|$ contains the intersection of the sets $|St\,A_i|, \cdots, |St\,A_k|$. On the other hand $|St\,\sigma|$ is contained in each of these sets. Therefore it is their intersection.

10. Let now φ be a mapping of the polyhedron $|K|$ into the polyhedron $|L|$. We shall say that K, L are *star-related relative to* φ whenever every set $\varphi|St\,A_i|$ is contained in some star-set $|St\,B_j|$. Suppose that this condition is fulfilled and select for each A_i a vertex $B_{j(i)}$ such that $\varphi|St\,A_i| \subset |St\,B_{j(i)}|$. Let f be the vertex transformation $\{A_i\} \to \{B_j\}$ defined by $fA_i = B_{j(i)}$. Take any $\sigma = A_i \cdots A_k \in K$. The sets $|St\,A_i|, \cdots, |St\,A_k|$ intersect and their intersection is $|St\,\sigma|$. Hence each set $\varphi|St\,A_i|, \cdots, \varphi|St\,A_k|$ contains $\varphi|St\,\sigma|$ and so they intersect. Since they are respectively contained in $|St\,B_{j(i)}|, \cdots, |St\,B_{j(k)}|$, these last sets also intersect. This implies that $\zeta = B_{j(i)} \cdots B_{j(k)} \in L$ and also that the intersection is $|St\,\zeta|$. The existence of ζ proves that f has a barycentric extension $K \to L$. Still denoting this extension for convenience by f, we will have $f\sigma = \zeta$. We refer to f as a *star-projection* $K \to L$.

Since each star-set $|St\,B_{j(i)}|, \cdots, |St\,B_{j(k)}|$ contains $\varphi|St\,\sigma|$, we have $\varphi|St\,\sigma| \subset |St\,\zeta|$. Hence if $x \in \sigma$, then φx and $y = fx$ are both in the closure $\bar{\zeta}_1$ of the simplex ζ_1 of $St\,\zeta$ containing φx. Therefore fx and φx may be joined by a segment in $|\bar{\zeta}_1|$ and hence in $|L|$. Referring then to (I, 15.4) we have proved the

(10.1) **Lemma.** *Let K, L be star-related to a mapping $\varphi : |K| \to |L|$. Then φ is homotopic to a barycentric mapping $K \to L$, where the homotopy paths are segments or points each contained in the closure of a simplex of L.*

Borders. Suppose that $|K|$ and $|L|$ have a nonvoid intersection Ω. The borders of Ω in each complex are certain closed neighborhoods defined as follows. For every σ of K let $\mu(\sigma)$ be the intersection of the closed set $|\overline{St\ \sigma}|$ with Ω. The border of $|L|$ in K, written $B(|L|)$, is the set of all the simplexes σ which have nonvoid sets $\mu(\sigma)$. (The designation $B(|L|)$ is somewhat incomplete but will not cause major difficulties.) If σ' is a face of σ, $\mu(\sigma') \supset \mu(\sigma)$. Hence if $\mu(\sigma)$ is nonvoid, so is $\mu(\sigma')$, and consequently if σ is in $B(|L|)$ so is every face of σ. Thus $B(|L|)$ is a subcomplex of K.

Noteworthy special case: $|L| \subset |K|$. Then the border is the closure of a particularly simple neighborhood of $|L|$ in $|K|$. This was the only case discussed before (II, 8, second case). In particular, in connection with the Jordan curve theorem $|L|$ was the Jordan curve J and K a subdivision of the two-sphere S containing J. The borders of J in appropriate subdivisions of K played an essential role in the proof of the theorem.

Refinement of complexes. This concept as given here is more general than as defined for sets (I, 8). However, the added degree of generality will be found useful in an important argument later.

Let the notations and the situation be as before. We will then say that K is a *refinement* of L, or *refines* L, whenever every nonvoid set $\mu(A_i)$ corresponding to a vertex A_i of K is contained in some set $|St\ B_j|$ corresponding to a vertex B_j of L.

In the special case where $|K| \subset |L|$, we have $B(|L|) = K$, and we weaken the definition to: K refines L if every $|St\ A_i|$ is contained in some $|St\ B_j|$. This is the usual relation of refinement between the collections of star-sets $\{|St\ A_i|\}$ and $\{|St\ B_j|\}$.

Let us suppose, without further restriction, that K refines L. For each nonvoid $\mu(A_i)$ choose a $B_{j(i)}$ such that $|St\ B_{j(i)}| \supset \mu(A_i)$. It is an elementary matter to show that $A_i \to B_{j(i)}$ defines a simplicial set-transformation and related barycentric mapping $\varphi : B\ (|L|) \to L$.

Noteworthy special case: $|K| \subset |L|$. Then φ is merely a star-projection relative to the identity on $|K|$ considered as a mapping $|K| \to |L|$.

(10.2) *A subdivision K_1 of K is a refinement of K.*

In fact, the verification of the star condition is immediate for $K_1 = K'$ and hence for any subdivision. The related star-projection is merely an inverse d^{-1} of chain-subdivision $d : K \to K_1$.

11. Suppose that φ is a mapping $|K| \to |L|$ but that K, L are not star-related as to φ. Let λ be the Lebesgue number of the covering $\mathfrak{B} = \{|St\ B_j|\}$ of $|L|$. Since φ is continuous and $|L|$, $|K|$ are compacta there is an $\eta > 0$ such that if C is any subset of $|K|$ and diam $C < \eta$, then diam $\varphi C < \lambda$, and hence C is in some $|St\ B_j|$. Choose now a subdivision K_1 of K whose mesh $< \frac{1}{2}\eta$. If $\{A_{1i}\}$ are the vertices of K_1, then diam $|St\ A_{1i}| < \eta$, and hence $\varphi\ |St\ A_{1i}|$ is in some $|St\ B_j|$. In other words K_1, L are star-related as to φ. Consequently φ is homotopic to a barycentric mapping $K_1 \to L$ with paths in closures of simplexes of L.

The same argument may be applied with L replaced by a subdivision L_1 of arbitrarily small mesh. Therefore we have:

(11.1) **Theorem.** *Every mapping φ of a polyhedron $|K|$ into a polyhedron $|L|$ is ϵ homotopic to a barycentric mapping $K_1 \to L_1$ where K_1, L_1 are suitable subdivisions of K,L. More precisely mesh $L_1 < \epsilon$ and the homotopy paths are each contained in the closure of a simplex of L_1.* (Alexander [c]).

A convenient result to have is the following property whose proof, entirely elementary, is left to the reader:

(11.2) *Let K, L, M be geometric complexes, where K and L are star-related as to a mapping $\varphi : |K| \to |L|$ with a star-projection f, likewise L and M as to a mapping $\psi : |L| \to |M|$ with a star-projection g. Then K, M are star-related relative to the mapping $\psi\varphi : |K| \to |M|$ with the star-projection gf.*

12. Application to certain dimensional invariance theorems. The propositions which we have in view refer to the invariance of the dimension of Euclidean spaces and regions and are due to Brouwer. The dimension alluded to is in fact the same as the dimension in the sense of Menger-Urysohn. In the proofs however it stands for the arithmetical dimensions meant when one says: n-cell, n-space, etc. The proofs are believed to be closely related to those given by Sperner on the basis of his lemma.

(12.1) **Lemma.** *If the polyhedron $|K|$ is contained in the polyhedron $|L|$ then* $\dim K \leqq \dim L$.

Choose a subdivision L_1 of L which is at the same time a refinement of K and let $B(|K|)$ be the border of K in L_1 (see 10) and φ the related star-projection $B(|K|) \to K$. Choose next a subdivision K_1 of K which refines L_1 and let ψ be the star-projection $K_1 \to L_1$. Since $|K_1| \subset |B(|K|)|$ this is actually a star-projection $K_1 \to B(|K|)$. Since $\varphi\psi$ is a star-projection $K_1 \to K$ it is an inverse d^{-1} of chain-subdivision $d : K \to K_1$.

Let $\dim K = p$, $\dim L = n$ and suppose the lemma incorrect or $p > n$. Thus K contains a σ^p and $\varphi\psi d\sigma^p = d^{-1} \cdot d\sigma^p = \sigma^p \neq 0$. On the other hand $\psi d\sigma^p \subset L_1$ and since $\dim L_1 = \dim L = n < p$, $\psi d\sigma^p = 0$, hence $\varphi\psi d\sigma^p = 0$. This contradiction proves the lemma.

(12.2) **Theorem.** *All the covering complexes of a given polyhedron Π have the same dimension. Its value is by definition the dimension of Π. This dimension is a topological invariant of the polyhedron.*

This is an obvious consequence of the lemma. This proposition is sometimes formulated as follows: The dimension of K is a topological invariant.

(12.3) **Theorem.** *If two Euclidean domains Ω, Ω' are topologically equivalent their spaces have equal dimensions.*

For if dim $\Omega = p$, dim $\Omega' = n$ and $p > n$, then some $\bar{\sigma}^n \subset \Omega$ would contain a $\bar{\sigma}^p \subset \Omega'$ which contradicts (12.2).

(12.4) **Corollary.** *A region of an Euclidean n-space cannot be parametrized by less than n-parameters.*

§5. The Brouwer Degree.

13. A closed orientable surface Φ is mapped by f on a two-sphere S. The degree ρ of f is the "algebraic" number of times that the image of the surface covers the sphere. If one thinks of the image as wrapped over the sphere, then one may conceivably smooth it out so that it never folds back and thus covers each point the same number of times which counted with a suitable sign is the degree ρ. If the covering has the orientation of the sphere ρ is positive, otherwise ρ is negative. An example of such a mapping is to multiply the longitude of each point by ρ.

Mappings on other surfaces than the sphere may also be considered. Thus the orthogonal projection of an Euclidean sphere on a plane is of degree zero.

The extension of the intuitive concept of degree to all dimensions offers no difficulty. For the sake of simplicity, and because it is the most interesting case, the treatment will mainly be confined to mappings of spheres on spheres.

It is instructive to examine first the degree for mappings of circumferences on circumferences. There is all the more reason for doing so since these mappings present certain minor exceptions which are best discussed separately.

Let then C, D be two circumferences respectively parametrized and oriented by means of two real variables u, v taken mod 1. A mapping $f : C \rightarrow D$ is determined by a relation $v = F(u)$, where F is continuous and $\rho = F(u + 1) - F(u)$ is a fixed integer independent of u. The number ρ is the degree of f. If f is slightly modified ρ will not change. Hence ρ is the same for homotopic mappings. Furthermore consider the particular mapping $g : v = \rho u$. If L is the real line of the variable v we may draw in L a segment joining ρu to $F(u)$, and this segment is unique when points v and $v + \rho$ are identified. It follows that the images of the segments on D generate a homotopy cylinder for f and g. Hence $f \backsim g$. Since g depends solely upon the degree we see that two mappings $C \to D$ are homotopic when and only when they have the same degree. In other words the homotopy classes of mappings $C \to D$ are in one-one correspondence with the degree. We will find that this is always the case for mappings of n-spheres on n-spheres.

There is an interesting application of the preceding considerations to the order of the points of a plane Π relative to a Jordan curve J in Π. Let P be a point of $\Pi - J$ and C a circumference of center P. Join P to a point Q of J and let the ray PQ meet C in R. Thus $Q \to R$ defines a mapping $F : J \to C$. Let C be oriented positively relative to Π. It is then found that the degree ρ of f is 0 for all exterior points and $+1$, or else -1, for all interior points. One may then orient J so that the degree $\geqq 0$. Under the circumstances its value $\omega(P)$ is the *order* of P

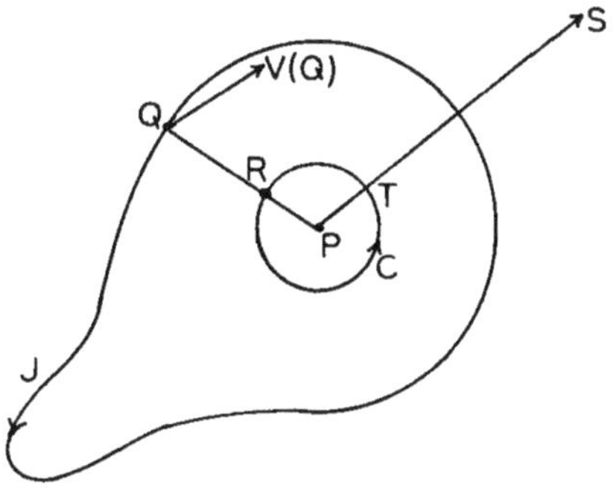

Figure 56

relative to J, and we have: $\omega(P) = 0$ at exterior points and $\omega(P) = 1$ at interior points. A number of proofs of the Jordan curve theorem have been based upon this simple property of the order.

We cannot forego another application. Let $\mathfrak{F}$ be a plane vector field defined at all points of the Jordan curve J and let the vector $V(Q)$ of $\mathfrak{F}$ vanish for no point Q of J. This time draw through P a ray PS parallel to $V(Q)$ and meeting the circumference C at T. Then $Q \to T$ defines a mapping $g : J \to C$ whose degree ρ is known as the *index* of J relative to the field. This index plays an all important role in the study, initiated by Poincaré, of the behavior of solutions of differential equations in the large (see *Survey*).

The interpretation of order and index in terms of angular variation is immediate. Both concepts are also very readily extended to higher dimensions.

The preceding examples will suffice to provide a good background for the discussion of the degree. Our general plan is to treat at length the degree for mappings of spheres on spheres. After that a few indications will suffice for the degree of mappings of circuits on circuits.

14. Let then S^n, Σ^n be two oriented n-spheres and f a mapping $S^n \to \Sigma^n$. Let $K = \{\sigma\}$ be a covering complex of S^n which is a $\mathfrak{B}\sigma^{n+1}$, i.e. isomorphic with the boundary complex of an $(n + 1)$-simplex. Let $L = \{\zeta\}$ be a similar complex for Σ^n. We assume the σ_i^n, ζ_i^n so oriented that

$$\gamma^n = \Sigma\sigma_i^n, \qquad \delta^n = \Sigma\zeta_i^n$$

are fundamental cycles for K, L.

Let now K_1 be a subdivision of K with related chain-subdivision d and set $\gamma_1^n = d\gamma^n = \Sigma d\sigma_i^n$. Since γ^n is a cycle of K and d is a chain-mapping, γ_1^n is a cycle of K_1.

We shall require later and may as well prove now:

$(14.1)_n$ $\gamma_1^n = d\gamma^n$ *is a fundamental n-cycle for* K_1. *That is to say every n-cycle of* K_1 *is of the form* $g\gamma_1^n$. *This implies the possibility of orienting* K_1 *by associating with it one of the two cycles* $\pm \gamma_1^n$. *If* K *is oriented by* γ^n *we shall agree to orient* K_1 *by* $d\gamma^n$ *and refer to this orientation as concordant with that of* K.

Since K_1 is merely a derived it is sufficient to prove (14.1) for the first derived K'. Moreover when this is done $(14.1)_1$ is elementary so we assume $(14.1)_{n-1}$, $n > 1$, and prove $(14.1)_n$.

Now any n-simplex of K' in our standard designation is of the form $\hat{\sigma}_i^0 \cdots \hat{\sigma}_j^n$, $\sigma_i^0 < \cdots < \sigma_j^n$. Hence any n-chain of K' is of the form

$$\gamma'^n = \Sigma C_i^{n-1}\hat{\sigma}_i^n, \qquad C_i^{n-1} \subset (\mathfrak{B}\sigma_i^n)',$$

and it is a cycle when and only when

$$F\gamma'^n = \pm \Sigma(FC_i^{n-1})\hat{\sigma}_i^n \pm \Sigma C_i^{n-1} = 0,$$

where the signs are unspecified since they are immaterial. Since no simplex of a C_i^{n-1} contains a vertex $\hat{\sigma}_i^n$, we must have $FC_i^{n-1} = 0$. Thus C_i^{n-1} is an $(n - 1)$-cycle of $(\mathfrak{B}\sigma_i^n)'$. Under the hypothesis of the induction $C_i^{n-1} = d\gamma_i^{n-1}$ where γ_i^{n-1} is a cycle for $\mathfrak{B}\sigma_i^n$. Now such a cycle is of the form $\lambda_i F\sigma_i^n$. Hence

$$\gamma'^n = \Sigma\lambda_i dF\sigma_i^n\hat{\sigma}_i^n = (-1)^n d\bar{\gamma}^n, \qquad \bar{\gamma}^n = \Sigma\lambda_i\sigma_i^n.$$

Since γ'^n is a cycle so is $d^{-1}\gamma'^n = d^{-1}d\bar{\gamma}^n = \bar{\gamma}^n$. Since γ^n is fundamental for K, $\bar{\gamma}^n = \lambda\gamma^n$, hence $\gamma'^n = \lambda(d\gamma^n)$. This proves (14.1).

Let the subdivision K_1 be selected of mesh so small that K_1 and L are star-related relative to the given mapping $f : S^n \to \Sigma^n$ and let φ be an associated barycentric mapping $K_1 \to L$ with induced chain-mapping still written φ. Consider now any oriented simplex ζ_j^n of L. Among the simplexes σ_i^n there will be say p_j such that $\varphi\sigma_i^n = +\zeta_j^n$ and q_j such that $\varphi\sigma_i^n = -\zeta_j^n$. The difference $\rho = p_j - q_j$ will be proved independent of ζ_j^n and is Brouwer's definition of the *degree of the mapping f.*

When $n = 1$ it is not difficult to identify the present degree with the one defined in (13). In the notations loc. cit. let K_1 and L be subdivisions of C, D star-related relative to f and let φ be the resulting star-projection. If $a_1, \cdots, a_r$ and $b_1, \cdots, b_s$ are the vertices of C, D as found consecutively in the proper order, then we will have $\varphi a_i = b_{j(i)}$, where $b_{j(i)}$, $b_{j(i+1)}$ are consecutive or coincident. An elementary consideration shows then that the degree in the sense of (13) is merely the algebraic number of times that the images of segments $a_i a_{i+1}$ cover a given $b_j b_{j+1}$, i.e. it is the degree in the sense just defined.

We shall now prove Brouwer's fundamental theorem on the degree:

(14.2) **Theorem.** *The degree ρ of the mapping f of an n-sphere S^n on another Σ^n depends solely upon f and it is the same for any mapping homotopic to f. In other words ρ depends solely upon the homotopy class of f.*

The mapping f is called *sense preserving* [*sense reversing*] when the degree ρ is positive [negative].

The variable elements in the definition of the degree ρ are ζ_j^n, φ, K_1, K, L and we must show above all that ρ does not depend upon them.

15. Since φ is a chain-mapping $\varphi\gamma^n$ is a cycle of L. Since L is n-cyclic with fundamental n-cycle δ^n, we must have $\varphi\gamma^n = \rho_1\delta^n$. Let $\sigma_{ji}'^n$, $\sigma_{ji}''^n$ be the n-simplexes of K_1 mapped respectively into $+\zeta_j^n$ and $-\zeta_j^n$. Thus

$$\gamma^n = \Sigma\sigma_{ji}'^n + \Sigma\sigma_{ji}''^n + C^n$$

where $\varphi C^n = 0$. Hence

$$\varphi\gamma^n = \rho_1\delta^n = \sum_j(\sum_i\varphi\sigma_{ji}'^n + \Sigma\varphi\sigma_{jh}''^n) = \Sigma(p_j - q_j)\zeta_j^n = \rho_1\Sigma\zeta_j^n.$$

Hence $p_j - q_j = \rho_1 = \rho$. Thus ρ_1 is the degree and it is clearly independent of ζ_j^n. In fact, we have found for it the following interpretation: ρ is the coefficient of δ^n in $\varphi\gamma^n$. This justifies in a measure the sense preserving or reversing terminology introduced above.

Let $\{A_i\}$ and $\{B_j\}$ be the vertices of K_1 and L. By hypothesis $f|St\,A_i|$ is contained in some star-set $|St\,B_j|$ and φ is determined by choosing for each A_i a vertex $B_j = \varphi A_i$ related to A_i in the above manner. Thus φ need not be unique and we must show that ρ is independent of the choice of φ.

It is clear that we merely need to show that ρ is not affected when the value of φ is changed on a single vertex A of K_1. Thus let $\varphi A = B$, $\varphi' A = B' \neq B$ and let $\varphi' = \varphi$ on all vertices other than A. Let now $\sigma^n = A\sigma^{n-1}$ be a nondegenerate simplex of K_1 with A as vertex. Thus σ^{n-1} does not have the vertex A and hence $\zeta = \varphi\sigma^{n-1} = \varphi'\sigma^{n-1}$. We do not mark the dimension of ζ since it may be degenerate. By hypothesis $f|St\, A\sigma^{n-1}|$ is contained in both star-sets $|St\, B\zeta|$, $|St\, B'\zeta|$. Hence the two sets have a nonvoid intersection and consequently $BB'\zeta$ is a simplex of L. Notice that

$$\text{(15.1)} \qquad \varphi A\sigma^{n-1} = B\zeta, \qquad \varphi' A\sigma^{n-1} = B'\zeta.$$

In addition we shall show that as a relation between n-chains

$$\text{(15.2)} \qquad B'\zeta - B\zeta + BB'F\zeta = 0.$$

If $\dim \zeta < n - 1$ every term is zero. If $\dim \zeta = n - 1$ then $BB'\zeta$ is an $(n + 1)$-simplex of an n-complex and hence degenerate. Hence $FBB'\zeta$, which is the left-hand side in (15.2), must be zero. We write (15.2) in the form

$$\text{(15.3)} \qquad B'\varphi\sigma^{n-1} - B\varphi\sigma^{n-1} + BB'\varphi F\sigma^{n-1} = 0.$$

If AC^{n-1} is a chain of $St\, A$, (15.3) may be applied to each simplex of the star and it yields

$$\text{(15.4)} \qquad B'\varphi C^{n-1} - B\varphi C^{n-1} + BB'\varphi FC^{n-1} = 0.$$

Consider in particular the chain AC^{n-1} sum of the oriented n-simplexes of K_1 with A as vertex. Thus $\gamma_1^n = AC^{n-1} + D^n$, where no simplex of D^n has A for vertex. We have

$$0 = F\gamma_1^n = C^{n-1} - AFC^{n-1} + FD^n.$$

At the right the middle term includes all the simplexes with A for vertex. Hence $FC^{n-1} = 0$. Hence from (15.4): $B'\varphi C^{n-1} = B\varphi C^{n-1}$ or $\varphi' AC^{n-1} = \varphi AC^{n-1}$. Since $\varphi' D^n = \varphi D^n$ we find $\varphi'\gamma_1^n = \varphi\gamma_1^n$. Therefore φ' yields the same degree ρ as φ. In other words the degree does not depend upon the special choice of φ.

We shall now free the degree from the choice of K_1. If K_1, K_2 are two subdivisions of K they have a common subdivision K_3 refining both. Hence it is sufficient to show that a subdivision K_2 of K_1 yields the same ρ as K_1 itself.

Let d_1 be chain-subdivision $K_1 \to K_2$, and d_1^{-1} an inverse of d_1. Then φd_1^{-1} is a suitable φ for K_2. The fundamental cycle in K_2 being $d_1 d\gamma^n = d_1\gamma_1^n$, the degree ρ' as defined by K_2 is governed by $\varphi d_1^{-1}(d_1\gamma_1^n)$

$= \rho'\delta^n$. Since $\varphi d_1^{-1} \times d_1 = \varphi$, and $\varphi\gamma_1^n = \rho\delta^n$ we have $\rho' = \rho$. Thus changing the subdivision K_1 of K does not affect the degree.

We have already shown that once K, L are given the degree ρ of f is uniquely determined. It is still necessary to free ρ from its dependence upon the two complexes.

Let first L_1 be a subdivision of L with d' as the related chain-subdivision. Let K_1 be a subdivision of K such that K_1, L_1 are star-related as to f and let ψ be the related star-projection $K_1 \to L_1$. According to (14.1) (with L, L_1 in place of K, K_1) $\psi\gamma_1^n = \rho_1(d'\delta^n)$. Remembering $(14.1)_n$ we see that $d'\delta^n$ is fundamental for L_1. Hence ρ_1 is the degree determined by means of the pair K_1, L_1 after the manner of the beginning of this number. Now $d'^{-1}\psi\gamma_1^n = \rho_1\delta^n$. But $d'^{-1}\psi$ is a star-projection $K_1 \to L$ and hence $\rho_1 = \rho$. Thus in the determination of the degree L may be replaced by L_1.

Consider now a new covering complex L^* of Σ^n which is still a simplex-boundary. Take a refinement L_1^* of L and L^* and a K_1 which is star-related to L_1^* and hence to L, as to f. Let δ_1^n be the fundamental cycle of L_1^*, ψ a star-projection $L_1^* \to L$, ω the star-projection $K_1 \to L_1^*$. Thus $\psi\omega = \varphi$ is a star-projection $K_1 \to L$. Now $\psi\delta_1^n = \lambda\delta^n$, and $\omega\gamma_1^n = \rho_1\delta_1^n$, where ρ_1 is the degree relative to L_1^*. Hence $\varphi\gamma_1^n = \lambda\rho_1\delta^n$ and therefore $\lambda\rho_1 = \rho$. Similarly $\rho_1 = \mu\rho$. Hence if one of ρ, ρ_1 is zero so is the other. If they are not then $\lambda\mu = 1$. Let us agree to orient L^*, hence L, i.e. to choose δ_1^n, so that $\lambda = 1$. Then $\lambda = \mu = 1$ and $\rho_1 = \rho$.

Suppose finally that K is replaced by an analogue K^*. This time we take a subdivision K_1^* of K^* which is also a refinement of K_1 where, as before, K_1 is star-related to L relative to f. Let ψ be the star-projection $K_1^* \to K_1$ and γ_1^{*n} the fundamental n-cycle of K_1^*. Here $\psi\gamma_1^{*n} = \nu\gamma_1^n$, $\varphi\gamma_1^n = \rho\delta^n$, hence $\varphi\psi\gamma_1^{*n} = \nu\rho\delta^n$. Since $\varphi\psi$ is a star-projection $K_1^* \to L$, $\nu\rho = \rho_1$ the degree defined by means of K^*. Assuming K^* oriented so that $\nu = 1$, we find again $\rho_1 = \rho$. Thus the degree is freed from the special choice of K, L provided that any similar pair K^*, L^* is oriented in a certain manner. This point will be discussed in a moment.

16. All that remains to be done is to show that the degree of f depends solely upon the homotopy class of f. Assuming then $f_o \backsim f_1$ we must show that f_o and f_1 have the same degree.

By hypothesis there is a mapping $\Phi : l \times S^n \to \Sigma^n$, $l : 0 \leqq u \leqq 1$, which agrees with f_o on $0 \times S^n$ and with f_1 on $1 \times S^n$. Let $f(u) = \Phi|u \times S^n$ and $G_i(u) = \Phi|u \times |St\, A_i|$. Denote also by λ the Lebesgue number of the finite open covering $\{|St\, B_j|\}$ of Σ^n.

Take now any u_o and let $l(\epsilon)$ denote the set $|u - u_o| < \epsilon$. Since Φ is continuous we may choose ϵ so small that every set $\Phi(l(\epsilon) \times |St\, A_i|)$

is of diameter $< \lambda$. Hence corresponding to A_i we may choose B_j such that $\Phi(l(\epsilon) \times |St\, A_i|) \subset |St\, B_j|$. Thus for every $u \in l(\epsilon)$ we will have $G_i(u) \subset |St\, B_j|$. Therefore for every such u we may define φ by $\varphi A_i = B_j$. Thus φ will be fixed, and hence the degree of $f(u)$ will be the same for all $u \in l(\epsilon)$. Since the segment l may be covered with a finite set of sets $\{l(\epsilon_1), \cdots, l(\epsilon_r)\}$ in which consecutive elements overlap, the degree is constant on l. Hence it is the same for f_o and f_1.

This completes the proof of Theorem (14.2).

17. Complements. (17.1) *The degree of the identity mapping* $S^n \to S^n$ *is unity. Hence the degree of a deformation* $S^n \to S^n$ *is likewise unity.*

The mutual orientations specified above for the two covering complexes K,K^* may be said to have been so chosen that the degree of the identity $S^n \to S^n$ as determined from K,K^* (the latter plays here the same role as L for Σ^n) is unity. Similarly for the mutual orientations of L,L^*.

A mapping $f : S^n \to \Sigma^n$ is inessential if it is homotopic to a constant mapping g (mapping of S^n into a point of Σ^n). Since g has evidently the degree zero we find:

(17.2) *An inessential mapping has the degree zero.*

(17.3) *The degree of a product of mappings of n-spheres into n-spheres is equal to the product of the degrees of the factors.*

Let S_1, S_2, S_3 be three n-spheres, f_i a mapping $S_{i+1} \to S_i$ and ρ_i its degree for $i = 2,1$. Cover S_i with a subdivision K_i of a simplex-boundary $\mathfrak{B}\,\sigma^{n+1}$ such that S_{i+1} and S_i are star-related relative to f_i and let φ_i be the induced chain-mapping $K_{i+1} \to K_i$. If γ_i^n is the fundamental n-cycle of K_i then $\varphi_i\gamma_{i+1}^n = \rho_i\gamma_i^n$. Hence $\varphi_1\varphi_2\gamma_3^n = \rho_1\rho_2\gamma_1^n$. Thus f_1f_2 has the asserted degree $\rho_1\rho_2$.

18. Extension to disks. Let Δ^n, Δ_1^n be two n-disks of dimension $n > 1$ and let S^{n-1}, S_1^{n-1} be their boundary spheres. The case $n = 1$ will not be required and does not fit very well into the general situation. We shall only consider mappings $f : \Delta_1^n \to \Delta^n$ such that $fS_1^{n-1} \subset S^{n-1}$ and briefly refer to them as *regular*. If f is regular then $f_1 = f|S_1^{n-1}$ is a mapping $S_1^{n-1} \to S^{n-1}$ and hence it has a degree r. This degree is by definition the *degree* of f.

Let f, g be two regular mappings $\Delta_1^n \to \Delta^n$, and let them be homotopic in the following sense. If l is the segment $0 \leqq u \leqq 1$ then there is a mapping $\Phi : l \times \Delta_1^n \to \Delta^n$ which agrees with f on $0 \times \Delta_1^n$, with g on $1 \times \Delta_1^n$ and is such that $\Phi(l \times S_1^{n-1}) \subset S^{n-1}$. In other words if $f_1 = f|S_1^{n-1}$, $g_1 = g|S_1^{n-1}$ and $\Phi_1 = \Phi|(l \times S_1^{n-1})$ then Φ_1 agrees with f_1 on $0 \times S_1^{n-1}$ and with g_1 on $1 \times S_1^{n-1}$. We thus have a homotopy of the mappings f_1, g_1 also. Under the circumstances we say that the homotopy determined by Φ is *regular* and also that it is an *extension* of the homotopy determined by Φ_1. We shall now prove:

(18.1) *A homotopy of f_1, g_1 may always be extended to a regular homotopy of the regular mappings f, g.*

It is convenient to take as representatives of each disk the interior and boundary of a sphere of radius one in an Euclidean n-space.

Suppose first $f_1 = g_1$. If $P_1 \in \Delta_1^n$ and $P = fP_1$, $P' = gP_1$ then either $P = P'$ or else there is a segment PP' in Δ^n. Hence f and g are regularly homotopic.

Consider now the general case. Let O, O_1 be the centers of Δ, Δ_1. Define a mapping f' as follows: $f'O_1 = O$; if $P_1 \in \Delta_1^n - O_1$, the ray O_1P_1 meets S_1^{n-1} in Q_1 and we let $Q = fQ_1 \in S^{n-1}$ and take as $f'P_1$ the point P of OQ such that $OP = O_1P_1$. Since $f|S_1^{n-1} = f'|S_1^{n-1}$, the mappings f and f' are regularly homotopic.

Let g' be deduced from g like f' from f. It is clear that in the proof of (18.1) we may substitute f', g' for f, g. Define then Φ as follows: $\Phi(l \times O_1) = O$; if P_1 is a point of Δ_1^n other than O_1 and the ray O_1P_1 meets S_1^{n-1} in Q_1, then $\Phi(u \times P_1)$ is at the intersection of the sphere of center O and radius O_1P_1 in Δ^n with the ray from O to $\Phi_1(u \times Q_1)$. This homotopy is manifestly regular.

Mappings of n-disks on n-spheres. The situation remaining the same, suppose that there exists a mapping g of Δ_1^n into a given n-sphere S^n such that gS_1^{n-1} is a fixed point A of S^n and that g is a homeomorphism $\Delta_1^n - S_1^{n-1} \to S^n - A$. If f is a regular mapping $\Delta^n \to \Delta_1^n$ then $h = gf$ is a mapping $\Delta^n \to S^n$ which we shall also refer to as *regular*. This term is not particularly appropriate since the property which it covers depends upon both Δ_1^n and A, but this will actually be the situation where such mappings are used later.

Let us define the *degree* ρ of f as also the degree of h. Let also a *regular homotopy* of h and h_1 designate a homotopy Φ such that $\Phi(l \times S^{n-1}) = A$. Finally let h be called *inessential* if it is regularly homotopic to h_1 such that $h_1\Delta^n = A$. Under the circumstances:

(18.2) *Properties* (17.2) *and* (17.3) *hold for regular mappings and homotopy* $\Delta^n \to S^n$.

Extension to n-circuits. All the properties of the degree for mappings of spheres on spheres may be extended automatically to mappings of orientable n-circuits on one another. This is quite clear since only circuit properties were utilized in our treatment.

§6. HOPF'S CLASSIFICATION OF MAPPINGS OF n-SPHERES ON n-SPHERES.

19. $(19.1)_n$ **Theorem.** *A n.a.c.s. for the homotopy of two mappings f, g of the n-sphere S^n into the n-sphere Σ^n is that they have the same degree. In other words the degree completely characterizes the homotopy classes of mappings of n-spheres into one another.* (Hopf [a]).

Necessity has already been established as part of Theorem (14.2). We assume then that f, g have the same degree ρ and will show that they are homotopic.

Theorem $(19.1)_0$ is obvious and $(19.1)_1$ has already been proved in substance in (14). We assume therefore $(19.1)_{n-1}$, $n > 1$, and prove $(19.1)_n$.

It will be convenient to have before us the following corollaries of $(19.1)_n$:

$(19.2)_n$. *A n.a.s.c. for $f: S^n \to \Sigma^n$ to be inessential is that its degree be zero.*

$(19.3)_{n+1}$. *A n.a.s.c. for regular homotopy of two regular mappings of an $(n+1)$-disk Δ_1^{n+1} into another Δ^{n+1} is that they have the same degree. Hence a n.a.s.c. for the regular mapping $f: \Delta_1^{n+1} \to \Delta^{n+1}$ to be inessential is that its degree be zero.*

A very simple observation will enable us to normalize the mappings $S^n \to \Sigma^n$. Suppose that α is a deformation $S^n \to S^n$ and β a deformation $\Sigma^n \to \Sigma^n$. Then as mappings $S^n \to \Sigma^n$, f and $\beta f \alpha$ are homotopic. Hence $f \backsim g$ is equivalent to $\beta f \alpha \backsim \beta g \alpha$. We shall take constant advantage of this property in the proof.

Let f, K_1, L be as in (14). Since the related star-projection $K_1 \to L$ is homotopic to f, we may assume that f is that star-projection. For convenience we also denote by f the induced chain-mapping.

Let us suppose explicitly that $L = \mathfrak{B}(A\zeta^n)$, $\zeta^n = B_0 \cdots B_n$. At the same time assume that Σ^n is Euclidean and that the antipode A' of A in Σ^n is in ζ^n and such that an arc of great circle from A to A' meets $\Sigma^{n-1} = |\mathfrak{B}\zeta^n|$ in one and only one point.

Take in particular any point P of Σ^{n-1} and let APA' be the arc of great circle from A to A' through P. Define a mapping β of Σ^n into itself as follows: $\beta AP = \beta A = A$; if $Q \in PA'$ then $R = \beta Q$ divides the arc APA' in the same ratio as Q divides PA'. The existence of the arc QR whenever $Q \neq R$ shows that β is a deformation.

Suppose that f is a mapping $S^n \to \Sigma^n$. Then $\beta f \backsim f$ maps into A all the points which f does not map on ζ^n.

Let us call *significant part* of f the set $f^{-1}\bar{\zeta}^n$. Suppose that f and g

have the same significant part which consists of a finite set of n-disks $\{\eta_i^n\}$ such that: (a) distinct disks η_i^n and η_j^n meet only, if they do, in boundary points; (b) if $f_i = f|\eta_i^n$, $g_i = g|\eta_i^n$, f_i and g_i are regular mappings $\eta_i^n \to \bar{\zeta}^n$. Under these conditions, which will be realized below, we prove:

(19.4) *A sufficient condition for $f \backsim g$ is $f_i \backsim g_i$ regularly, for every i.*

This property will greatly simplify the comparison of f and g. Consider the cylinders $\Pi = l \times S^n$, $\Pi_i = l \times \eta_i^n$, $l : 0 \leqq u \leqq 1$. By hypothesis there is a mapping $\Phi_i : \Pi_i \to \bar{\zeta}^n$ which agrees with f_i on $0 \times \eta_i^n$, with g_i on $1 \times \eta_i^n$ and maps $l \times |\mathfrak{B}\eta_i^n|$ on Σ^{n-1}. Hence $\beta\Phi_i$ agrees with βf_i on $0 \times \eta_i^n$, with βg_i on $1 \times \eta_i^n$ and maps $l \times |\mathfrak{B}\eta_i^n|$ on A. Define now $\Phi : \Pi \to \Sigma^n$ as follows: $\Phi|(l \times \eta_i^n) = \beta\Phi_i$; for all points P of Σ^n not on any η_i^n, $\Phi(l \times P) = A$. Since $\Phi|(l \times |\mathfrak{B}\eta_i^n|) = A$, Φ is a mapping. Since it agrees with βf on $O \times S^n$ and with βg on $1 \times S^n$, we have $\beta f \backsim \beta g$ and hence $f \backsim g$ by (19.3).

A noteworthy consequence of the result just proved is

(19.5) *When f_i is inessential, η_i^n may be suppressed from the essential part of f.*

For let f_i be replaced by $f_i' : \eta_i^n \to A$. As a consequence f is replaced by a new mapping f' which according to (19.4) $\backsim f$, and η_i^n is not in the significant part of f'.

(19.6) Our next step will consist in reducing the essential part of f to a standard form of the type just considered and depending solely upon the degree ρ. Consider the n-simplexes $\{\sigma_i^n\}$ of K_1 mapped by f onto ζ^n and suppose that the first p are mapped on $+\zeta^n$, and the other in number q on $-\zeta^n$, so that $p - q = \rho$.

Let us fix our attention on a given σ_i^n of the set just considered and let S_i^{n-1} be its boundary sphere oriented by the cycle $F\sigma_i^n$. Let η_i^n be a small convex n-disk or "island" in σ_i^n, P a nonboundary, point of η_i^n and let the boundary sphere $S_i'^{n-1}$ of the disk be so oriented that its projection from P onto S_i^{n-1} has the degree unity. Since the projection is in a fixed homotopy class the orientation is independent of P and we refer to this orientation as "concordant with that of S^n."

Now a ray PM from P to any point M of S_i^{n-1} will meet $S_i'^{n-1}$ in a single point M'. Define a mapping $\alpha : S^n \to S^n$ as follows: On $S^n - \sigma_i^n$ the mapping is the identity; $\alpha M'M = M$; PM' is mapped barycentrically on PM.

Let now f be replaced by $f\alpha$. Since α is topological on $S_i'^{n-1}$, $f\alpha$ maps $S_i'^{n-1}$ with the same degree on $|\mathfrak{B}\zeta^n|$ as f maps S_i^{n-1}. Hence $f\alpha$ maps η_i^n on the n-disk $\bar{\zeta}^n$ with the same sign as f maps σ_i^n on $\bar{\zeta}^n$. This process will be called *concentrating f on η_i^n in σ_i^n*. Let this concentration be operated simultaneously on all the σ_i^n. Its effect will be to reduce the significant part of f to a finite set of disjoint islands $\{\eta_i^n\}$ of which the first p are mapped on $\bar{\zeta}^n$ with degree $+1$ and the q others with degree -1. As before $p - q = \rho$ is the degree of f.

Since the concentration may be made on any η_i^n in σ_i^n, η_i^n may be displaced at will in σ_i^n. On the other hand if η_i^n is small enough we may always surround it with a small simplex σ'^n not necessarily in K_1 and displace η_i^n within σ'^n. Hence we may choose η_i^n so small that we may freely displace it on S^n provided that in the process it never touches any similar island resulting from the other σ_i^n.

Let us suppose that neither p nor q are zero and let η_i^n, η_j^n be two islands mapped on $\bar{\zeta}^n$ with opposite degrees. Let us displace the two, as we may, so that they are in two n-simplexes $C\sigma^{n-1}$, $-D\sigma^{n-1}$ of S^n. In such a position the orientations of the two simplexes will be concordant with those of the two islands. Hence reversing our process (which merely causes a homotopy in f) we may assume that f maps the disks η^n, η'^n determined by the two simplexes on $\bar{\zeta}^n$ with opposite degrees. Now $\eta^n \cup \eta'^n = \eta''^n$ is an n-disk which has for boundary sphere $|\mathfrak{B}C\sigma^{n-1}| \cup |\mathfrak{B}D\sigma^{n-1}| - \sigma^{n-1}$ and for fundamental n-cycle $F(C\sigma^{n-1}) + F(-D\sigma^{n-1})$. Hence $f|\eta''^n$ is regular and of degree zero, and so it is inessential. It follows that η''^n may be suppressed from the significant part of f (19.5). The total effect on f is thus the reduction of p and q by one unit. This process may be carried out until one of these numbers is zero.

To sum up then, we may reduce the significant part to $|\rho|$ islands $\{\eta_i^n\}$ mapped on $\bar{\zeta}^n$ with the same degree $+1$ or -1.

The same method of displacement will enable us to replace the islands by as many preassigned simplexes. In short we may reduce f to a mapping f_1 whose significant part consists of ρ simplexes $\{\sigma_i^n\}$ each mapped by f_1 on ζ^n with the same sign as ρ.

Let then f and g have the same degree ρ and let f_1, g_1 be the reduced mappings in the sense just described. By hypothesis $(19.1)_{n-1}$ holds and hence $(19.3)_n$ holds likewise. Since $f_1|\sigma_i^n$ and $g_1|\sigma_i^n$ have the same degree we may replace the first, by means of a homotopy regular in $\bar{\sigma}_i^n$, by a mapping which agrees with the second in $\bar{\sigma}_i^n$. Operating likewise on every $\bar{\sigma}_i^n$ we shall have brought f and g to agreement on all the $\bar{\sigma}_i^n$. Hence (19.4) $f \backsim g$ and Theorem (19.1) is proved.

§7. Some Theorems on the Sphere.

20. The material in the present section has been communicated to the author by A. W. Tucker [b]. A certain algebraic lemma resembling Sperner's lemma is first established and various known theorems are then deduced from the lemma.

Let $x_1, \cdots, x_{n+1}$ be coordinates for an Euclidean space $\mathfrak{E}^{n+1}$. The set Σ^n

$$|x_1| + \cdots + |x_{n+1}| = 1$$

is the boundary of the convex region obtained when $=$ is replaced by $\leq$. Hence Σ^n is an n-sphere. This representation of the n-sphere turns out to be particularly convenient for our purpose. Let $X_{\pm i}$ denote the unit-points on the x_i-axis and consider any simplex

$$\sigma^p = X_{i_0} \cdots X_{i_p}, \; |i_0| < \cdots < |i_p|.$$

The collection $\{\sigma\}$ is manifestly a complex K covering Σ^n. Let K_1 be any subdivision (barycentric derived) of K and let δ be chain-subdivision $K \to K_1$. Let also A denote the symmetry of the space $\mathfrak{E}^{n+1}$ relative to the origin, i.e. the operation sending the point with coordinates x_i into the point with coordinates $-x_i$. As an operation on both K and K_1 it is a (topological) barycentric mapping of order two. Let us also denote by A the induced simplicial chain-mappings in K and in K_1. Two elements of any sort corresponding under A are called *antipodal.*

We shall only use chains and cycles mod 2 and this fact must be kept in mind throughout.

(20.1) **Lemma.** *Let f be a barycentric mapping $K_1 \to K$ sending antipodal vertices of K_1 into antipodal vertices of K. Then each n-simplex of K is the image of an odd number of n-simplexes of K_1. Hence the degree of f is odd.*

For convenience we also denote by f the chain-mapping which it induces.

The first step towards this lemma is to set up a system of $n + 1$ pairs of antipodal chains on K

$$b^n, Ab^n; b^{n-1}, Ab^{n-1}; \cdots ; b^o, Ab^o \tag{20.2}$$

such that $b^n + Ab^n$ is the sum of all n-simplexes of K, $b^o + Ab^o$ is a single pair of antipodal vertices, and

$$Fb^p = b^{p-1} + Ab^{p-1} = FAb^p \tag{20.3}$$

$$(p = n, n-1, \cdots, 1).$$

To accomplish this let

$$b^p = \text{the sum of all } p\text{-simplexes } X_{\pm 1} X_{\pm 2} \cdots X_{\pm p} X_{p+1},$$

$$Ab^p = \text{the sum of all } p\text{-simplexes } X_{\pm 1} X_{\pm 2} \cdots X_{\pm p} X_{-p-1},$$

for $p = n, n - 1, \cdots, 0$. Then, clearly, $b^n + Ab^n$ includes all n-simplexes of K, $b^o + Ab^o = X_1 + X_{-1}$, and (20.3) holds.

The second step towards the lemma is to demonstrate that, if $c^k, Ac^k; c^{k-1}, Ac^{k-1}; \cdots; c^o, Ac^o$ are chains of K such that

$$Fc^p = c^{p-1} + Ac^{p-1} = FAc^p \tag{20.4}$$

$$(p = k, k - 1, \cdots, 1)$$

and $c^o + Ac^o$ is a single pair of antipodal vertices, then

$$c^k + Ac^k \neq F(d^{k+1} + Ad^{k+1}) \tag{20.5}$$

for any pair of antipodal $(k + 1)$-chains d^k, Ad^k (possibly both zero). The proof is by induction, as follows. If $k = 0$, (20.5) holds because the left side is a single pair of antipodal vertices and the right side is the sum of an even number of pairs of antipodal vertices. Now suppose (20.5) already proved for $k = 1, 2, \cdots, p - 1$ and assume

$$c^p + Ac^p = F(d^{p+1} + Ad^{p+1}).$$

Then, transposing terms and letting A commute with F,

$$c^p + Fd^{p+1} = A(c^p + Fd^{p+1}).$$

Due to this invariance with respect to A, $c^p + Fd^{p+1}$ can be expressed as the sum of two antipodal p-chains, say

d^p = the sum of those p-simplexes whose last vertex has a positive subscript;

Ad^p = the sum of those p-simplexes whose last vertex has a negative subscript.

That is,

$$c^p + Fd^{p+1} = d^p + Ad^p.$$

Operating with F, substituting from formula (20.5), and using $FF = 0$,

$$c^{p-1} + Ac^{p-1} = F(d^p + Ad^p).$$

But this is contrary to the preceding stage in the induction. Therefore formula (20.5) holds for $k = p$ also. By induction it holds for all k. In particular for $k = n$, it shows that

$$c^n + Ac^n \neq 0. \tag{20.6}$$

Let

$$c^p = f\delta b^p \qquad (p = n, n - 1, \cdots, 0)$$

where b^p comes from (20.2). Since A commutes by hypothesis with f and naturally with δ,

$$Ac^p = f\delta(Ab^p).$$

Since F commutes with the mappings f and δ, and A has period 2, (20.4) above is satisfied for $p = n, n - 1, \cdots, 1$. Hence (20.5) holds. But

$$F(c^n + Ac^n) = 0. \tag{20.7}$$

The relations (20.6) and (20.7) together imply that $c^n + Ac^n$ is the sum of all the n-simplexes of K. For one can pass from any one n-simplex σ_1^n of K to any other σ_2^n by a sequence of moves in which the sign of one vertex is changed at a time. If σ_1^n belongs to $c^n + Ac^n$ and σ_2^n did not, there would be some stage in the sequence where two n-simplexes, one belonging to $c^n + Ac^n$ and the other not, would have an $(n - 1)$-face in common. But such an $(n - 1)$-face would appear in the boundary of $c^n + Ac^n$, contrary to (20.7).

The same conclusion that $c^n + Ac^n$ includes all the n-simplexes of K may also be reached as follows: It is easily shown that K is an n-circuit. Hence the non-trivial n-cycle mod 2, $c^n + Ac^n$, is the fundamental n-cycle taken mod 2, that is to say the sum mod 2 of all the n-simplexes of K.

In conclusion,

$$b^n + Ab^n = \text{the sum of all } n\text{-simplexes of } K;$$

$$\delta(b^n + Ab^n) = \text{the sum of all } n\text{-simplexes of } K_1;$$

$$f\delta(b^n + Ab^n) = c^n + Ac^n = \text{the sum of all } n\text{-simplexes of } K.$$

Therefore each n-simplex of K is the image under f of an odd number of n-simplexes of K_1.

21. Let S^n denote the n-sphere

$$x_1^2 + x_2^2 + \cdots + x_{n+1}^2 = 1.$$

A homeomorphism between S^n and Σ^n is immediately established by central projection from the origin. Under this homeomorphism antipodal points on Σ^n correspond to antipodal points on S^n.

(21.1) **Theorem.** *The intersection of $n + 1$ closed sets on S^n is not empty, if the $n + 1$ sets and their $n + 1$ antipodal sets together cover S_n, but no one of the sets contains a pair of antipodal points.*

Transfer the closed sets from S^n to Σ^n by central projection. Let the resulting covering of Σ^n by $n + 1$ pairs of antipodal closed sets be denoted by

$$G_1, AG_1; G_2, AG_2; \cdots ; G_{n+1}, AG_{n+1}.$$

Choose a subdivision K_1 whose mesh is less than the Lebesgue number of this covering. Associate each vertex of K_1 with the first set of the covering, as ordered above, which contains the vertex. Then put the sets of the covering, as ordered above, in one-one correspondence with the vertices of K, ordered as follows

$$X_1, X_{-1}; X_2, X_{-2}; \cdot \cdot \cdot ; X_{n+1}, X_{-n-1}.$$

As a result each vertex of K_1 is mapped into a vertex of K so that antipodal vertices of K_1 are mapped into antipodal vertices of K and adjacent vertices of K_1 (i.e. joined by an edge) are mapped into nonantipodal vertices of K. Hence the mapping is simplicial and satisfies the hypothesis of the preceding lemma. By the lemma the n-simplex $X_1 X_2 \cdot \cdot \cdot X_{n+1}$ is the image of an odd number of n-simplexes of K_1. Any one of the n-simplexes of K_1 which is mapped into it has one vertex in each of the sets $G_1, G_2, \cdot \cdot \cdot , G_{n+1}$. Therefore the intersection of these $n + 1$ sets is not empty since the mesh of K_1 is less than the Lebesgue number of the covering.

(21.2) **Theorem.** *If S^n is covered by $n + 1$ closed sets, at least one of the sets contains a pair of antipodal points.* (Lusternik-Schnirelmann, 1930).

Suppose S^n is covered by $n + 1$ closed sets, no one of which contains a pair of antipodal points. The conditions of the preceding theorem are met. Therefore the $n + 1$ sets have some point in common. But the point antipodal to this common point is covered by at least one of the sets. This contradiction proves the theorem.

(21.3) **Theorem.** *If n closed sets and their n antipodal sets together cover S^n, then at least one of the n closed sets contains a pair of antipodal points.*

Suppose S^n is covered by the n pairs of antipodal closed sets

$$G_1, AG_1; G_2, AG_2; \cdot \cdot \cdot ; G_n, AG_n,$$

no one of which contains a pair of antipodal points. Adjoin $G_{n+1} = AG_n$, $AG_{n+1} = G_n$ to get a covering by $n + 1$ pairs of antipodal closed sets, no one of which contains a pair of antipodal points. Then by the first theorem the intersection of $G_1, G_2, \cdot \cdot \cdot , G_{n+1}$ is not empty. But this is impossible, since G_n and $G_{n+1} = AG_n$ are disjoint or else G_n contains some pair of antipodal points.

(21.4) **Theorem.** *There is no (continuous) mapping of S^n into S^{n-1} which maps antipodal points of S^n into antipodal points of S^{n-1} without exception.*

Suppose

$$y_\alpha = f_\alpha(x_1, x_2, \cdots, x_{n+1}), \qquad \alpha = 1, 2, \cdots, n$$

is a (continuous) mapping of S^n

$$x_1^2 + x_2^2 + \cdots + x_{n+1}^2 = 1$$

into S^{n-1}

$$y_1^2 + y_2^2 + \cdots + y_n^2 = 1$$

such that

$$f_\alpha(-x_1, -x_2, \cdots, -x_{n+1}) = -f_\alpha(x_1, x_2, \cdots, x_{n+1}),$$
$$\alpha = 1, 2, \cdots, n.$$

Form the n pairs of antipodal closed sets on S^n determined by the inequalities

$$\left.\begin{array}{l} G_\alpha : f_\alpha(x_1, x_2, \cdots, x_{n+1}) \geqq \dfrac{1}{\sqrt{n}}, \\ AG_\alpha : f_\alpha(x_1, x_2, \cdots, x_{n+1}) \leqq \dfrac{-1}{\sqrt{n}}, \end{array}\right\} \alpha = 1, 2, \cdots, n.$$

These cover S^n since $|y_\alpha|$ is not less than $1/\sqrt{n}$ for all coordinates of a point on S^{n-1}. No G_α contains a pair of antipodal points since G_α and AG_α are disjoint. Therefore the assumed mapping (f_α) cannot exist.

(21.5) **Theorem.** *A (continuous) mapping of S^n into Cartesian n-space maps some pair of antipodal points into a single point.* (Borsuk-Ulam, 1933).

Suppose

$$y_\alpha = f_\alpha(x_1, x_2, \cdots, x_{n+1}), \qquad \alpha = 1, 2, \cdots, n$$

is a (continuous) mapping of S^n into Cartesian n-space $(y_1, y_2, \cdots, y_n)$ such that

$$f_\alpha(-x_1, -x_2, \cdots, -x_{n+1}) \neq f_\alpha(x_1, x_2, \cdots, x_{n+1}),$$
$$\alpha = 1, 2, \cdots, n.$$

Then

$$\rho(x_1, x_2, \cdots, x_{n+1}) = \Sigma\{f_\alpha(x_1, x_2, \cdots, x_{n+1}) - f_\alpha(-x_1, -x_2, \cdots, -x_{n+1})\}^{1/2}$$

does not vanish on S^n. The mapping

$$y_\alpha = \{f_\alpha(x_1, x_2, \cdots, x_{n+1}) - f_\alpha(-x_1, -x_2, \cdots, -x_{n-1})\}/\rho(x_1, x_2, \cdots, x_{n+1})$$

is a (continuous) mapping of S^n into S^{n-1} which maps antipodal points into antipodal points. But this contradicts the previous theorem. Therefore no mapping (f_α) as supposed can exist.

PROBLEMS

(For all but the first two problems the author is indebted to A. W. Tucker.)

1. Let $\Pi = \{E_i^p\}$ be a polyhedron whose cells E_i^p are convex polyhedral. Consider chains $C^p = \Sigma g_i E_i^p$ and show that the boundary operator F may be defined so as to have $FF = 0$. Let Π still denote the resulting *"generalized complex."* Prove that the barycentric derived Π' may be defined as when K is simplicial, with related chain-mappings δ, τ such as in (5) satisfying $\tau\delta = 1$.

2. A subdivision K_1 of an n-circuit K is an n-circuit and if one of K, K_1 is orientable so is the other.

3. Show that an n-cube can be divided into $n!$ n-simplexes as follows. Represent the n-cube in Cartesian n-space by

$$0 < y_i < 1 \qquad i = 1, 2, \cdots, n;$$

for each permutation $i_1, i_2, \cdots, i_n$ of $1, 2, \cdots, n$, form the open subset determined by

$$0 < y_{i_1} < y_{i_2} < \cdots < y_{i_n} < 1;$$

and parametrize this open subset by

$$x_1 = y_{i_1}$$

$$x_\alpha = y_{i\alpha} - y_{i\alpha-1} \qquad \alpha = 2, \cdots, n$$

$$x_{n+1} = 1 - y_{i_n}.$$

4. Show that a centrally-symmetric simplicial subdivision of an n-cube can be obtained as follows. Represent the n-cube by

$$|y_i| < 1 \qquad i = 1, \cdots, n;$$

partition the n-cube into 2^n subcubes by means of the coordinate hyperplanes $y_i = 0$ $(i = 1, \cdots, n)$; and divide each subcube into $n!$ n-simplexes determined by

$$0 < |y_{i_1}| < |y_{i_2}| < \cdots < |y_{i_n}| < 1.$$

5. By central projection from the origin in Cartesian $(N + 1)$-space let the closed N-simplex determined by

$$\Sigma x_i = 1, \qquad 0 \leq x_i \leq 1 \qquad i = 1, 2, \cdots, N + 1$$

be put in *bicontinuous one-one* correspondence with the system of $N + 1$ adjoining N-faces, opposite to the origin, of the closed $(N + 1)$-cube

$$0 \leq x_i \leq 1 \qquad i = 1, 2, \cdots, N + 1.$$

Then the closed N-simplex can be subdivided into $N + 1$ convex N-cells, each the projective image of an N-cube.

a. Show that barycentric subdivision of the above closed N-simplex induces a simplicial subdivision of the system of $N + 1$ adjoining N-cubes.

b. Use the above methods to obtain an arbitrarily fine simplicial subdivision of a (finite) n-complex.

6. Let K be the complex resulting from the subdivision of the closed n-cell: $|x_1| + \cdots + |x_n| \leq 1$ by the coordinate hyperplanes $x_i = 0$. Let K_1 be any centrally symmetric subdivision of K. Then it is not possible to label each vertex of K_1 with one of the $2n$ number $\pm 1, \cdots, \pm n$ so that: (a) no two vertices joined by a one-simplex received complementary labels, i.e. labels such as $+i$, $-i$; (b) every two antipodal vertices on the boundary $\Sigma|x_i| = 1$ receive complementary labels.

7. There exists no mapping of the n-disk into the $(n - 1)$-sphere sending each pair of antipodal boundary-points into a pair of antipodal points.

8. Any mapping of an n-disk into an $(n - 1)$-sphere maps at least one pair of antipodal boundary-points into a single point.

9. If nonnull vectors are distributed continuously throughout the n-disk then there is at least one pair of antipodal boundary-points whose vectors are parallel.

10. Let $f_i(x) = f_i(x_1, \cdots, x_n)$ be continuous and real valued throughout the n-disk. If

$$\frac{f_1(x)}{x_1} = \cdots = \frac{f_n(x)}{x_n} > 0$$

for no boundary-point x then the system $f_i = 0$, $(i = 1, \cdots, n)$ has at least one solution in the disk.

11. It is not possible to cover the n-disk by n pairs of closed sets so that the two sets of each pair are disjoint and no one of the $2n$ sets contains a pair of antipodal boundary-points.

12. In a covering of an n-disk by n closed sets at least one of the sets contains a pair of antipodal boundary-points.

V. Further Properties of Homotopy. Fixed Points. Fundamental Group. Homotopy Groups

In the present chapter chain-mappings are further developed and applied to various questions notably to subdivision and to the proof of the author's fixed point formula. The chapter contains also a rather complete discussion of Poincaré's fundamental group, as well as an introduction to the homotopy groups of Hurewicz.

§1. Homotopy of Chain-mappings.

1. Considerations of integration (see *Survey*) have given many indications of a close relationship between homotopy and homology. The exact connection will be shown to rest upon an algebraic analogue of point-set homotopy.

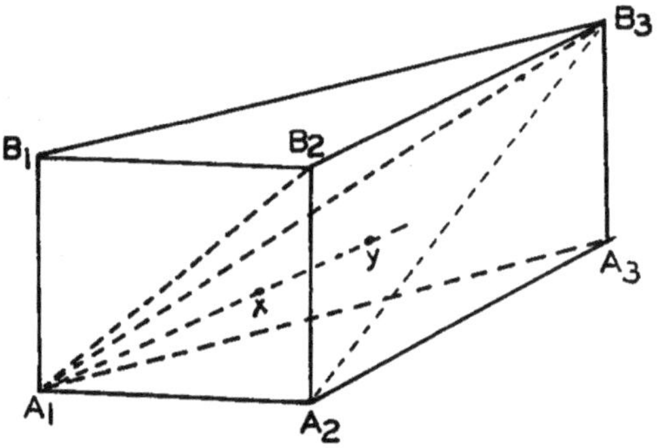

Figure 57

Let $K = \{\sigma\}$ be a geometric complex, l the segment $0 \leq u \leq 1$, and $\Pi = l \times |K|$ the product of the segment by $|K|$. We may manifestly assume the situation such that Π is a prism, in some Euclidean space $\mathfrak{E}^r$, built on $|K|$ with edge of length l. Let f denote the translation sending a point $0 \times x = x$ of $|K|$ into the point $1 \times x$. Clearly $0 \times \sigma \rightarrow 1 \times \sigma$ or $\sigma \rightarrow f\sigma$ defines an isomorphism of the "base" K with the other base which we shall denote by K^*. Let $\{A_i\}$ be the vertices of K and $B_i = fA_i$. Thus $\{B_i\}$ are the vertices of K^*. We suppose the A_i ranged in a certain order and write any simplex of K as $\sigma = A_i \cdots A_j$, $i < \cdots < j$. Similarly with K^* and the B_i. Corresponding to any $\sigma^p = A_i \cdots A_j \cdots A_h$ of K construct all the simplexes

$$(1.1) \qquad \zeta_j^{p+1} = A_i \cdots A_j B_j \cdots B_h.$$

We shall show that

(1.2) *The ζ's together with all their faces make up a simplicial complex $\mathfrak{K}$ covering* Π.

In Fig. 57 the construction is illustrated when K is a triangle. The resulting decomposition described in (1.2) is the well known decomposition of a triangular prism into three tetrahedra.

Since the cell $l \times \sigma^p$ is convex and all the vertices of ζ_j^{p+1} are vertices of that cell, the latter contains ζ_j^{p+1}. To prove that $\mathfrak{K}$ is a simplicial complex we must prove that: (a) the simplexes of $\mathfrak{K}$ are disjoint; (b) every point of Π is in one of them.

Let $x \in l \times \sigma^p$, $p > 0$. The ray from A_i to x meets $\sigma^{*p} \cup |l \times \mathfrak{B}\sigma^p|$ in a point y. Let $u(y)$ denote the value of u at y. If $u(y) = 1$ then $y \in \bar{\sigma}^{*p}$ and hence $x \in \overline{A_i\sigma^{*p}}$, i.e. x is in an $A_i\sigma^{*q}$, $\sigma^{*q} < \sigma^{*p}$, or $x = y$ and x is in a face of σ^{*p} itself. Suppose now $u(y) < 1$. If x were common to two distinct simplexes of $\mathfrak{K}$, y would have the same property relative to the simplexes of $\mathfrak{K}$ in $|l \times \mathfrak{B}\sigma^p|$. This reduces the proof of (a) for points in an $l \times \sigma^p$ to the same for points in an $l \times \sigma^q$, $q < p$. Since (a) is trivial for $x \in l \times \sigma^0$, (a) holds throughout. Similarly (b) for x reduces so the same for y, and as above to the same for $p = 0$, where it is also obvious.

2. Consider now the linear chain operator $\mathfrak{D} : K \to \mathfrak{K}$ defined by

$$(2.1) \qquad \mathfrak{D}A_{i_0} \cdots A_{i_p} = \Sigma(-1)^q A_{i_0} \cdots A_{i_q} B_{i_q} \cdots B_{i_p}.$$

An elementary calculation yields the boundary relation

$$(2.2) \qquad F\mathfrak{D}A_{i_0} \cdots A_{i_p} = B_{i_0} \cdots B_{i_p} - A_{i_0} \cdots A_{i_p} - \mathfrak{D}FA_{i_0} \cdots A_{i_p}.$$

If φ is the chain-mapping (isomorphism) induced by f then $\varphi A_i \cdots A_j = B_i \cdots B_j$. Hence (2.2) may be written

$$(2.3) \qquad F\mathfrak{D}\sigma^p = \varphi\sigma^p - \sigma^p - \mathfrak{D}F\sigma^p.$$

Since all the operations in (2.3) are linear we may replace σ^p by any chain C^p and thus obtain

$$F\mathfrak{D}C^p + \mathfrak{D}FC^p = \varphi C^p - C^p,$$

which is more conveniently written in operator form:

$$(2.4) \qquad F\mathfrak{D} + \mathfrak{D}F = \varphi - 1.$$

Suppose now that g is a barycentric mapping of $\mathfrak{K}$ into a new complex L and let τ_1 be the induced chain-mapping. Applying τ_1 to both sides of (2.4) we find since τ_1 commutes with F:

$$(2.5) \qquad F(\tau_1\mathfrak{D}) + (\tau_1\mathfrak{D})F = \tau_1\varphi - \tau_1.$$

Let us set $\tau_2 = \tau_1\varphi$, so that τ_2 like τ_1 is a chain-mapping $K \to L$. Let us also write $\mathfrak{D}$ for $\tau_1\mathfrak{D}$ so that $\mathfrak{D}$ is now also a linear operator $K \to L$. Under the circumstances (2.5) assumes the form

$$(2.6) \qquad F\mathfrak{D} + \mathfrak{D}F = \tau_2 - \tau_1.$$

This relation will form the basic connecting link between homotopy and homology. It may be said to contain the algebraic substance of homotopy. It justifies also the

Definition. Two chain-mappings $\tau_1, \tau_2 : K \to L$ are said to be *homotopic*, written $\tau_1 \backsim \tau_2$, whenever there exists a linear operator $\mathfrak{D} : K \to L$ such that (2.6) holds. Whenever $\tau_1 = 1$ the chain-mapping τ_2 is called a *deformation*.

Thus in view of (2.4) the chain-mapping φ is a deformation. This is quite natural since φ merely expresses the effect of a translation on the chains of the base K of the prism.

3. The fundamental property of homotopy in chains is:

(3.1) *Two homotopic chain-mappings $\tau_1,\tau_2 : K \to L$ induce the same homomorphisms of the homology groups of K into the corresponding groups of L.*

An equivalent formulation of (3.1) is: If γ^p is any cycle of K then $\tau_1\gamma^p$ and $\tau_2\gamma^p$ are in the same homology class of L, or $\tau_1\gamma^p \sim \tau_2\gamma^p$ in L. In fact if we apply both sides of (2.6) to γ^p and recollect that $F\gamma^p = 0$, we find $F\mathfrak{D}\gamma^p = \tau_2\gamma^p - \tau_1\gamma^p \sim 0$, as asserted.

Another convenient property is:

(3.2) *If τ_1, τ_2 are chain-mappings $K \to L$ and τ a chain-mapping $L \to M$, and if $\tau_1 \backsim \tau_2$ then $\tau\tau_1 \backsim \tau\tau_2$.*

For applying τ to both sides of (2.6) and since τ commutes with F,

$$F(\tau\mathfrak{D}) + (\tau\mathfrak{D})F = \tau\tau_2 - \tau\tau_1.$$

This proves the asserted homotopy and shows furthermore that its operator is $\tau\mathfrak{D}$.

Finally let us observe that

(3.3) *Homotopy of chains is an equivalence relation.*

For the relationship between τ_1, τ_2 in (2.6) is symmetric since it holds when τ_1, τ_2 are interchanged and $\mathfrak{D}$ is replaced by $-\mathfrak{D}$. It is reflexive since it holds for $\tau_1 = \tau_2$ and $\mathfrak{D} = 0$. As for transitivity let $\tau_1 \backsim \tau_2$, $\tau_2 \backsim \tau_3$ with respective operators $\mathfrak{D}_1$, $\mathfrak{D}_2$. Thus

$$\mathfrak{D}_1 F + F\mathfrak{D}_1 = \tau_2 - \tau_1, \; \mathfrak{D}_2 F + F\mathfrak{D}_2 = \tau_3 - \tau_2,$$

and therefore

$$(\mathfrak{D}_1 + \mathfrak{D}_2)F + F(\mathfrak{D}_1 + \mathfrak{D}_2) = \tau_3 - \tau_1.$$

Thus $\tau_3 \backsim \tau_1$ with operator $\mathfrak{D}_1 + \mathfrak{D}_2$. This proves (3.3).

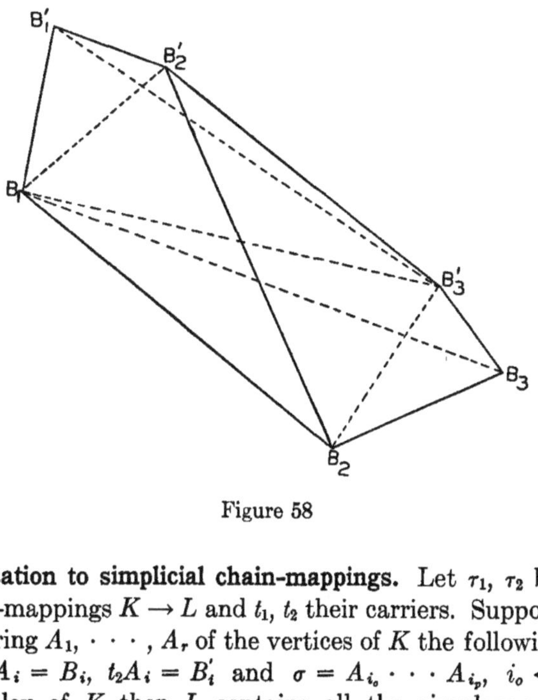

Figure 58

4. Application to simplicial chain-mappings. Let τ_1, τ_2 be two simplicial chain-mappings $K \to L$ and t_1, t_2 their carriers. Suppose that in a certain ordering $A_1, \cdots, A_r$ of the vertices of K the following property holds: If $t_1 A_i = B_i$, $t_2 A_i = B'_i$ and $\sigma = A_{i_0} \cdots A_{i_p}$, $i_0 < \cdots < i_p$ is any simplex of K then L contains all the simplexes $B_{i_0} \cdots B_{i_q} B'_{i_q} \cdots B'_{i_p}$.

In Fig. 58 the situation is described when σ is a triangle $A_1A_2A_3$. Our condition asserts then that L must contain the simplexes $B_1B_2B_3B'_3$, $B_1B_2B'_2B'_3$ and $B_1B'_1B'_2B'_3$.

Under the circumstances we propose to show that τ_1, τ_2 are homotopic with operator $\mathfrak{D}$ given by

$$\text{(4.1)} \qquad \mathfrak{D}A_{i_0} \cdots A_{i_p} = \Sigma(-1)^q B_{i_0} \cdots B_{i_q} B'_{i_q} \cdots B'_{i_p}.$$

In fact this relation is entirely analogous to (2.1) save that in the sum one may replace A by B and B by B', and similarly in (2.2). Degenerate simplexes may occur this time but they do not affect the calculation. The analogue of (2.2) yields now in place of (2.4) the relation (2.6). This proves our assertion: $\tau_1 \backsim \tau_2$.

Noteworthy special case: Suppose that whatever $\sigma \in K$ the vertices of $t_1\sigma$ and of $t_2\sigma$ make up together the vertices of a simplex ζ of L and thus $t\sigma = \zeta$ is a common carrier of τ_1, τ_2. The two chain-mappings τ_1, τ_2 are then said to be *prismatically related*. It is clear that in this case the condition on the carriers holds whatever the ordering of the vertices of K and therefore $\tau_1 \backsim \tau_2$.

5. Extension of chain-mappings. The extension properties which we discuss are useful in various questions, notably in connection with subdivision and the related operations δ, τ.

In the statements to follow there occur closed set-transformations $t : K \to L$ with the property that if $\sigma \in K$, $t\sigma$ is a zero-cyclic closed subcomplex of L. We shall then say that t is *zero-cyclic*.

If the homotopic mappings τ_1, $\tau_2 : K \to L$ have a common carrier t such that if $\mathfrak{D}$ is the homotopy operator $\mathfrak{D}x$ is a chain of tx, then τ_1, τ_2 are said to be *homotopic* in t, written $\tau_1 \backsim \tau_2$ in t.

We shall now prove two propositions due to A. W. Tucker. His initial theorems had somewhat larger scope, but the special cases here considered will suffice for our purpose.

(5.1) **Theorem.** *Let t be a zero-cyclic set-transformation $K \to L$. If τ_o is a chain-mapping $K^o \to L^o$ (the zero-sections) which preserves the Kronecker index of zero-chains and has the carrier $t_o = t|K^o$, τ_o may be extended to a mapping $\tau : K \to L$ with the carrier t.*

(5.2) *The same property holds if τ_o is defined on a closed subcomplex $K_1 \supset K^o$ and of course with the carrier $t|K_1$.*

(5.3) **Theorem.** *Let K,L,t be as in the preceding theorem. If τ_1, τ_2 are chain-mappings $K \to L$ with τ as common carrier, and if $KI(\tau_1C^o) = KI(\tau_2C^o)$ whatever C^o, then $\tau_1 \backsim \tau_2$ in t.*

Proof of (5.1). The proof will consist in extending τ to the successive sections K^p of K. If the extension has been made to K^p we shall say that $(5.1)_p$ holds.

By definition we require $\tau|K^o = t_o$ so that $(5.1)_o$ holds. By hypothesis if $\sigma_i^1 \epsilon K$ then $KI(\tau F\sigma_i^1) = KI(F\sigma_i^1) = 0$. Since t is closed and zero-cyclic, $\tau F\sigma_i^1$ is a chain of the zero-cyclic complex $t\sigma_i^1$ and since its index vanishes $t\sigma_i^1$ contains a chain ζ_i^1 such that $F\zeta_i^1 = \tau F\sigma_i^1$. Hence if we define for each σ_i^1, $\tau\sigma_i^1 = \zeta_i^1$ we will have $\tau F = F\tau$ in K^1 and $(5.1)_1$ will hold.

Suppose now that $(5.1)_{p-1}$, $p > 1$, holds and consider $\sigma_i^p \in K$. By hypothesis $\tau F \sigma_i^p$ is a cycle of dimension > 0 in the zero-cyclic complex $t\sigma_i^p$. Hence $t\sigma_i^p$ contains a chain ζ_i^p such that $F\zeta_i^p = \tau F \sigma_i^p$. Hence if we set $\tau\sigma_i^p = \zeta_i^p$ we will have $\tau F = F\tau$ in K^p and $(5.1)_p$ will hold. This proves (5.1).

The treatment of (5.2) is the same save that now only $\sigma_i^p \in K_1$ needs to be considered.

Proof of (5.3). We shall define $\mathfrak{D}$ throughout the successive sections K^p. If the extension has been made to K^p we shall say that $(5.3)_p$ holds.

Consider σ_i^o. By hypothesis $\tau_1\sigma_i^o$ and $\tau_2\sigma_i^o$ are both chains of the zero-cyclic complex $t\sigma_i^o$ whose Kronecker indices are equal. Since $t\sigma_i^o$ is a zero-cyclic complex $\tau_2\sigma_i^o - \tau_1\sigma_i^o = F\zeta_i^1$ where ζ_i^1 is a chain of $t\sigma_i^o$. If we set $\mathfrak{D}\sigma_i^o = \zeta_i^1$ we will have

$$\mathfrak{D}F + F\mathfrak{D} = \tau_2 - \tau_1 \tag{5.4}$$

as applied in K^o. Therefore $(5.3)_o$ will hold. Assume now that $(5.3)_{p-1}$, $p \geq 1$, holds. Since t is a common carrier of τ_1 and τ_2, $t\sigma_i^p$ contains $\tau_1\sigma_i^p$ and $\tau_2\sigma_i^p$. On the other hand since t is closed if $\sigma_j^{p-1} < \sigma_i^p$ then $t\sigma_j^{p-1} \subset t\sigma_i^p$ and by $(5.3)_{p-1} : \mathfrak{D}\sigma_j^{p-1} \subset t\sigma_j^{p-1} \subset t\sigma_i^p$. Therefore $\mathfrak{D}F\sigma_i^p \subset t\sigma_i^p$. Hence $t\sigma_i^p$ contains

$$\eta_i^p = (-\mathfrak{D}F + \tau_2 - \tau_1)\sigma_i^p.$$

Applying F to both sides and recalling $(5.3)_{p-1}$ we find

$$F\eta_i^p = (-F\mathfrak{D} + \tau_2 - \tau_1)F\sigma_i^p = \mathfrak{D}FF\sigma_i^p = 0.$$

Therefore η_i^p is a cycle of the zero-cyclic complex $t\sigma_i^p$. Since $p > 0$, η_i^p bounds in $t\sigma_i^p$ or $\eta_i^p = F\zeta_i^{p+1}$, $\zeta_i^{p+1} \subset t\sigma_i^p$. Hence defining $\mathfrak{D}\sigma_i^p = \zeta_i^{p+1}$ (5.4) will hold in K^p, i.e. $(5.3)_p$ will hold. This proves (5.3).

6. Application to subdivision. Let $K = \{\sigma\}$ be our usual complex and let K', δ, τ be the derived of K and related operations as in (IV, 5). We have already proved the basic relation

$$\tau\delta = 1. \tag{6.1}$$

Proceeding a step farther we shall now prove the homotopy

$$\delta\tau \backsimeq 1 \tag{6.2}$$

whose important consequences will soon be clear.

If $\zeta = \hat{\sigma}_i \cdots \hat{\sigma}_j$, $\sigma_i < \cdots < \sigma_j$ is a simplex of K', then $\tau\zeta$ consists of simplexes which are faces of σ_j, or $\tau\zeta \subset Cl\, \sigma_j$. Hence $\delta\tau\zeta \subset (Cl\, \sigma_j)'$. It follows that if we define $t\zeta = (Cl\, \sigma_j)'$, t is a carrier for $\delta\tau$. Since $\zeta \subset (Cl\, \sigma_j)'$, t is likewise a carrier for the identity. Since $(Cl\, \sigma_j)'$ is the closed join of $\hat{\sigma}_j$ and $(\mathfrak{B}\sigma_j)'$, $(Cl\, \sigma_j)'$ is zero-cyclic, (III, 11.2). Finally

τ, δ, hence also $\delta\tau$, each send a vertex into a vertex and so they preserve the Kronecker index. This is evidently true for the identity. All the conditions of the homotopy theorem (5.3) are thus fulfilled regarding $\delta\tau$ and the identity. Hence $\delta\tau \backsim 1$ in t.

Let $\bar{\delta}$, $\bar{\tau}$ denote the homomorphisms in the homology groups induced by δ, τ. From (6.1), (6.2) and (3.1) follows

$$(6.3) \qquad \bar{\tau}\bar{\delta} = 1, \qquad \bar{\delta}\bar{\tau} = 1.$$

Hence we may state:

(6.4) *$\bar{\delta}$, $\bar{\tau}$ are isomorphisms, each the inverse of the other. Hence K,K' have isomorphic homology groups.*

The same treatment may be applied in relation to $K^{(s)}$ and K save that this time $(Cl\ \sigma)^{(s)}$ is proved zero-cyclic by repetition of (6.4).

(6.5) **Theorem.** *Let K_1 be a subdivision of K, d and d^{-1} the related chain-subdivision and an inverse. Then d,d^{-1} induce isomorphisms $\bar{d},\bar{d}^{-1}$ of the corresponding homology groups of the two complexes, each the inverse of the other. We also note the relations*

$$(6.6) \qquad d^{-1}d = 1, \qquad dd^{-1} \backsim 1.$$

§2. HOMOLOGY IN POLYHEDRA. RELATION TO HOMOTOPY.

7. By identifying in a certain way the homology groups of the complexes covering a given polyhedron Π there will be obtained homology groups for Π which are topologically invariant. An important associated result is the topological character of the homology groups, hence also of the Betti numbers, of the covering complexes. The system which we shall develop is inspired by the Čech homology theory for topological spaces (1932) and also closely related to Alexander [c] (1926).

The following notations will greatly simplify the treatment: If K,L are geometric complexes star-related relative to some mapping, the associated star-projection $K \to L$ will be denoted by KL and the induced homomorphism of the homology groups by $\overline{KL}$. Schematically, as we have already done on occasion, especially if there are several projected pairs, we shall represent the complexes by points: $K,L, \cdots$, and the projections by directed arcs $KL, \cdots$.

It follows at once from the definitions that

(7.1) *$KL \cdot LM$ is a KM.*

While projections need not be unique we have:

(7.2) *If there are two projections KL, $(KL)'$ the induced homomorphism in the homology groups is the same for both. That is to say $\overline{KL} = \overline{(KL)'}$ is unique.*

As a consequence under the same conditions as in (7.1):

$$(7.3) \qquad \overline{KL} \cdot \overline{LM} = \overline{KM}.$$

Proof of (7.2). Let t,t' be the barycentric mappings which induce the projections. If $\sigma \in K$, $\sigma = A_i, \cdots, A_j$, let $tA_i = B_i$, $t'A_i = B'_i, \cdots$. Then by the definition of t,t' : $t\,|St\,A_i| \subset |St\,B_i|$, $t'\,|St\,A_i| \subset |St\,B'_i|, \cdots$. Hence $|St\,B_i|,|St\,B'_i|, \cdots$, intersect and therefore L contains a simplex $\zeta = B_i \cdots B'_j = (t\sigma)(t'\sigma)$. Thus t,t' are prismatically related and so (7.2) follows from (4).

Since we are primarily interested in homology the nonuniqueness of KL will cause no difficulty.

8. Suppose now that all the complexes cover the same polyhedron Π. If $\{A_i\}$, $\{B_j\}$ are the vertices of K,L the statement: K refines L means that every star-set $|St\,A_i|$ is contained in some star-set $|St\,B_j|$. The collection $\{|St\,B_j|\}$ is a finite open covering of Π and if λ is the Lebesgue number of this covering K refines L whenever mesh $K < \frac{1}{2}\lambda$, i.e. whenever mesh K is sufficiently small.

If K refines L we have the related star-projection KL and the induced homomorphism $\overline{KL}$ of the homology groups of K into the corresponding groups of L. We shall now prove the basic

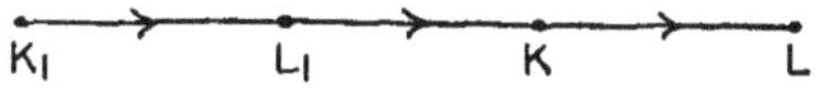

Figure 59

(8.1) **Theorem.** *The operation* $\overline{KL}$ *is an isomorphism.*

Let d denote chain-derivation in K and $\bar{d}$ the induced isomorphism and let us identify homology classes of K,K' which correspond under d; similarly for K', K'', etc. Thus all the homology groups $\mathfrak{H}^p(K^{(s)};G)$ for a given G, i.e. the groups in all the subdivisions of K, are now identified. Let the same thing be done for L. Choose now a subdivision L_1 of L then a subdivision K_1 of K of meshes so small that $\{K,L_1,K_1\}$ is an interlocking system in the sense of (II, 8). The corresponding diagram is shown in Fig. 59. As operations on the new groups chain-subdivisions (d in K, d' in L) reduce to she identity. Hence by (7.3):

$$\overline{K_1L_1} \cdot \overline{L_1K} = \bar{d}^{-1} = 1,$$

and similarly

$$\overline{L_1K} \cdot \overline{KL} = 1.$$

This is sufficient to prove that $\overline{KL}$ is an isomorphism and furthermore that $\overline{K_1L_1} = \overline{KL} = (\overline{L_1K})^{-1}$.

Suppose now that K,L are any two covering complexes. Choose any M refining both K and L. Then $\pi = \overline{ML} \cdot (\overline{MK})^{-1}$ is an isomorphism of the homology groups of K with the corresponding groups of L. Therefore:

(8.2) **Theorem.** *The corresponding homology groups of any two covering complexes K,L of a polyhedron Π are isomorphic. Hence they are topological invariants of Π.*

As an immediate corollary:

(8.3) *The Betti numbers and hence the characteristic of a covering complex of a polyhedron Π depend solely upon Π and thus they are topological invariants of the polyhedron.*

Notice also that (8.2) and (8.3) imply:

(8.4) *If $|K|$ and $|L|$ are homeomorphic the corresponding homology groups of K and L are isomorphic and they have equal Betti numbers.*

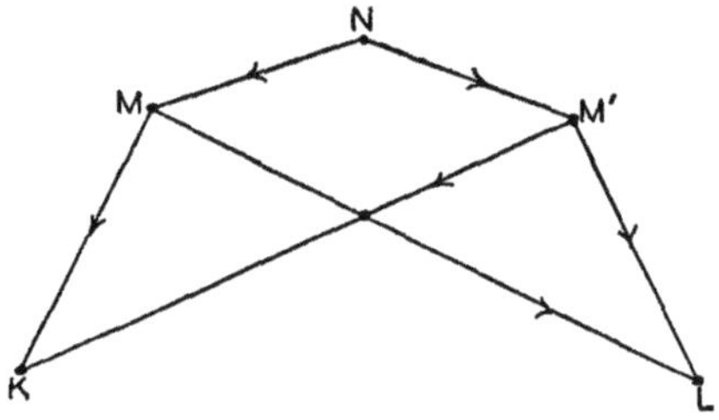

Figure 60

For L is isomorphic with a complex covering $|K|$.

The above noteworthy theorems were first proved, for rational Betti numbers, by J. W. Alexander [a], and later extended by Veblen [V] and Alexander [c]. Our proof is closely related to Alexander [c].

The significance of the preceding results becomes clear when one remembers that to calculate the Betti numbers, and in fact, all the homology groups, one merely requires a single covering complex.

The complete identification of all the groups $\mathfrak{H}^p(K;G)$ for all the K's covering Π is now quite simple. Let M',π' be a second pair like M,π. Let N refine both M and M'. The various complexes and operations are indicated in Fig. 60. The possible isomorphisms between any two complexes are represented by the oriented arcs joining them. For pairs such as (N,K), (N,L) these isomorphisms do not depend upon the paths. From this the same result is deduced for any pair. This proves our assertion.

Let $\pi(K,L)$ be the unique operation π thus obtained. Evidently $\pi(L,K) = \pi^{-1}(K,L)$. Hence by identifying elements of the homology groups of K,L corresponding under $\pi(K,L)$ and this for all pairs K,L, there arises an identification of all the groups $\mathfrak{H}^p(K;G)$ into a single group $\mathfrak{H}^p(\Pi,G)$, the p-th homology group of Π over G. It is clear that $\mathfrak{H}^p(\Pi;G) \cong \mathfrak{H}^p(K;G)$. The elements of these new groups are the *homology classes* of Π and the dimension $R^p_\pi(\Pi)$ of the groups mod π is the *p-th Betti number* of Π mod π. Here again $R^p_\pi(\Pi) = R^p_\pi(K)$. The "cyclic" terminology applies in the obvious way to Π.

The cycles of maximum dimension. Let $\dim \Pi = n$, hence $\dim K = n$ and let $\Gamma^n(\Pi)$ be a homology class of Π. Under the isomorphism of the homology groups there corresponds to $\Gamma^n(\Pi)$ a class $\Gamma^n(K)$ of K. Let γ^n be a cycle of $\Gamma^n(K)$. This cycle is unique. For if γ'^n were another in the same class we would have $\gamma^n - \gamma'^n \sim 0$, hence $\gamma^n - \gamma'^n = 0$.

Suppose in particular Π cyclic in the dimension n and let $\Gamma^n(\Pi)$ be a fundamental class (in the obvious sense similar to that of (III, 21)). One may show as done there that the only fundamental classes are $\pm\Gamma^n$ and there are the natural consequences relative to orientation.

As an application suppose that Π may be covered by two n-circuits $K = \{\sigma\}$ and $L = \{\zeta\}$, and suppose K orientable. Then K is cyclic in the dimension n and so is Π. Let the σ_i^n be so oriented that a fundamental cycle in K is given by $\gamma^n(K) = \Sigma\sigma_i^n$. This cycle determines $\Gamma^n(K)$, hence $\Gamma^n(\Pi)$ and consequently $\gamma^n(L)$ is fundamental for L. Therefore the ζ_i^n's suitably oriented will satisfy

$$\gamma^n(L) = \Sigma\zeta_i^n.$$

Thus the orientation in K determines one in L.

9. Let now Π,Ω be two polyhedra and f a mapping $\Pi \to \Omega$. We shall prove:

(9.1) **Theorem.** *The mapping f determines a homomorphism φ of the homology groups of Π into the corresponding groups of Ω and φ is the same for any two homotopic mappings of Π into Ω. In other words the homomorphism φ depends solely upon the homotopy class of f.*

Let $\{K\}$, $\{L\}$ be the covering complexes of Π,Ω. Corresponding to a given K choose an L, such that K,L are star-related relative to f and we have the related star-projection $KL : K \to L$ and induced operation in the homology groups $\overline{KL}$.

Under the identification of the homology groups of the K's and of the L's, $\overline{KL}$ becomes a homomorphism $\varphi(K,L)$ of the homology groups of Π into the corresponding groups of Ω. We first prove

(9.2) *$\varphi(K,L)$ depends solely upon f.*

That is to say φ is independent of the pair K,L which serves to define it.

We must show that if K_1,L_1 is another pair such as K,L then $\varphi(K_1,L_1) = \varphi(K,L)$. The proof is made very simple by means of a diagram. Select an L_2 refining L and L_1 then a K_2 refining K,K_1,L_2 so that the projections are as indicated in Fig. 61. In view of (7.3) the paths K_2KL and K_2L_2L represent the same homomorphism of the homology groups of K_2 into those of L. By reference to the groups of Π and Ω and since $\overline{K_2K}$, $\overline{L_2L}$ are isomorphisms we find that $\varphi(K,L) = \varphi(K_2,L_2)$. Similarly $\varphi(K_1,L_1) = \varphi(K_2,L_2)$, hence $\varphi(K,L) = \varphi(K_1,L_1)$, which is (9.2).

Thus $\varphi(K,L)$ is independent of K,L. We denote it by φ and it is the homomorphism described under (9.1).

Consider now two mappings $f,g : \Pi \to \Omega$ and let $\varphi(f)$, $\varphi(g)$ be the induced operations φ. We must prove:

(9.3) *If* $f \backsim g$ *then* $\varphi(f) = \varphi(g)$.

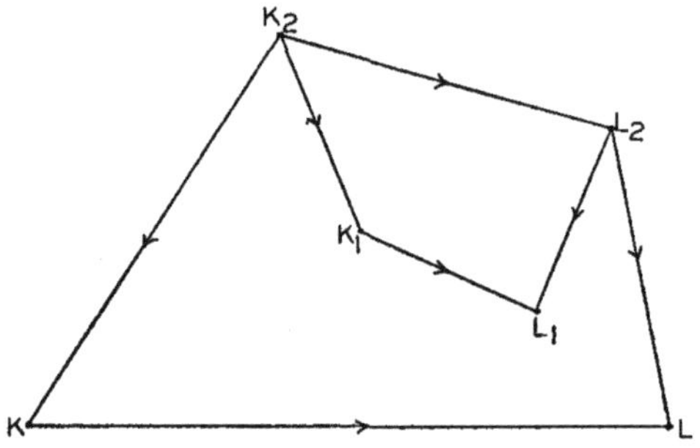

Figure 61

Let l be the segment $0 \leq u \leq 1$ and Φ a mapping $l \times \Pi \to \Omega$ which agrees with f on $0 \times \Pi$ and with g on $1 \times \Pi$. Let f_u denote the mapping $\Phi|u \times \Pi$. We have shown in (IV, 16) that corresponding to any u_o there is a subdivision K_1 of K such that K_1 and L are star-related relative to f_{u_o}. Referring now to the proof of (IV, 16) together with a very simple continuity argument, it is such that there is an $\epsilon > 0$ such that for $|u - u_o| < \epsilon$, K_1 and L are star-related relative to f_u and with a fixed induced chain-mapping $\pi : K_1 \to L$ for all u on the interval in question. This implies that $\varphi(f_u)$ is fixed on the interval. Since l may be covered with a finite set $\{l_i\}$ of such intervals where l_i, l_{i+1} overlap, $\varphi(f_u)$ is independent of u. Hence $\varphi(f) = \varphi(f_o) = \varphi(f_1) = \varphi(g)$.

Application to the degree. Suppose that Π and Ω are n-dimensional and with covering complexes K,L which are n-circuits. If $\Gamma^n(\Pi),\Gamma^n(\Omega)$ are the fundamental classes then the theorem asserts that $\varphi\Gamma^n(\Pi) = \rho\Gamma^n(\Omega)$. To determine ρ take a subdivision K_1 of K and let γ_1^n, δ^n be the fundamental n-cycles of K_1 and L. If K_1 is so chosen that K_1 and L are star-related relative to f and π is the induced star-projection then $\pi\gamma_1^n \sim \rho\delta^n$, hence $\pi\gamma_1^n = \rho\delta^n$ since n is the dimension of L. Thus we recognize in ρ the degree of f in the sense of Brouwer. The homotopy theorem has enabled us to define it directly so that its topological nature becomes obvious.

Application to deformation. Let us suppose $\Omega \subset \Pi$ and such that there are covering complexes L,K of Ω,Π with L a subcomplex of K. Let there exist also a deformation $f : \Pi \to \Omega$ which reduces to the identity in Ω. Then we shall prove:

(9.4) *The homomorphism $\varphi(f)$ induced by f in the homology groups is an isomorphism. Hence Π and Ω have isomorphic homology groups and in particular equal Betti numbers.*

Consider in fact a subdivision K_1 of K such that K_1,L are star-related relative to f. Let π be the induced star-projection $K_1 \to L$ and let d denote chain-subdivision $K \to K_1$. Thus πd is a chain-mapping $K \to L$. Since $\bar{d}$ is an isomorphism and $\bar{\pi} = \varphi(f)$, $\theta = \pi d$ induces $\bar{\theta} = \overline{\pi d} = \varphi(f)$. Thus we merely need to show that $\bar{\theta}$ is an isomorphism.

(a) *$\bar{\theta}$ is onto.* For if the cycle $\gamma^p \subset L$ then $\theta\gamma^p = \gamma^p$; hence $\bar{\theta}$ transforms the class of γ^p in K into its class in L.

(b) *$\bar{\theta}$ is univalent.* For suppose $\theta\gamma^p \sim 0$ in L. Then $\pi(d\gamma^p) = \theta\gamma^p \sim 0$ in K. Now d^{-1} is a chain-mapping $K_1 \to K$ induced by the identity $\Pi \to \Pi$ and π is one induced by a deformation f. Since $f \backsimeq 1$, π and d^{-1} induce the same homomorphism in the homology groups. Hence $0 \sim \theta\gamma^p = \pi(d\gamma^p) \sim d^{-1}(d\gamma^p) \sim \gamma^p$. Therefore $\bar{\theta}$ is univalent.

Properties (a) and (b) show that $\bar{\theta}$ is an isomorphism and prove (9.4).

Application: The homology groups of $\bar{\sigma}^p$ and of a tree are those of a point, i.e. both are zero-cyclic. For both are deformable in the manner here considered into a vertex.

§3. The Lefschetz Fixed Point Theorem for Polyhedra.

10. Under certain conditions it is possible to deduce from the associated equations of transformation of the homology groups the presence of fixed points in a mapping. This result, when stated with more precision, is the fixed point theorem. Its great generality and power will be shown by examples. We must first deal however with an analogous algebraic property for complexes.

(10.1) *The fixed element formula for complexes.* Let $K = \{\sigma_i^p\}$ be a complex and τ a chain-mapping $K \to K$ given by

$$\tau\sigma_i^p = \Sigma a_{ij}^p\sigma_j^p. \tag{10.2}$$

It is natural to consider the element σ_i^p as fixed if it enters on the right hand side of the relation, and then to assign to it the "weight" $(-1)^p a_{ii}^p$ as fixed element. Let us set $a^p = ||a_{ij}^p||$ and similarly for analogous matrices throughout. Let also

$$\Phi(\tau) = \Sigma(-1)^p \text{ trace } a^p, \tag{10.3}$$

where as usual

$$\text{trace } a^p = \Sigma a_{ii}^p$$

so that $(-1)^p$ trace a^p is the sum of the weights of the fixed p-simplexes. It is clear that $\Phi(\tau) \neq 0$ implies that some $a_{ii}^p \neq 0$ and therefore that τ has one or more fixed elements. This may be viewed as the general guiding principle in the whole question.

Consider now the effect of adopting new bases $\{x_i^p\}$ for the integral chain groups. We will have

$$\sigma_i^p = \Sigma \xi_{ij}^p \, x_j^p.$$

Since the transformation has an inverse of same form the determinant $|\xi^p| = \pm 1$, where $\xi^p = ||\xi_{ij}^p||$. The effect of τ on the x_i^p is now given by

$$\tau x_i^p = \Sigma b_{ij}^p \, x_j^p,$$

$$b^p = (\xi^p)^{-1} \, a^p \, \xi^p.$$

As is well known trace b^p = trace a^p, and therefore also

$$\Phi(\tau) = \Sigma(-1)^p \text{ trace } b^p.$$

Thus whatever the bases selected for the chains the form of $\Phi(\tau)$ remains the same. For this reason one may refer to $\Phi(\tau)$ as the *trace invariant of* τ.

All this holds of course for changes of bases mod π, which is what we shall require in a moment.

Example. τ is the identity. Then every $a^p = 1$, trace $a^p = \alpha^p$, the number of p-simplexes in K, and

$$\Phi(\tau) = \Sigma(-1)^p \alpha^p = \chi(K),$$

the Euler-Poincaré characteristic of K.

Let us pass now from the bases $\{\sigma_i^p\}$ for the chains to the rational bases $\{\beta_i^p, \gamma_j^p, \delta_k^p\}$ of (III, 16) for the rational cycles and let us write down the corresponding equations of τ. Since τ commutes with F it sends cycles into cycles and bounding cycles into bounding cycles. Therefore a $\tau\beta$ is a linear combination of bounding cycles β, and a $\tau\gamma$ one of cycles β, γ. Hence

$$(10.4)_p \qquad \tau\beta_i^p = \Sigma b_{ij}^p \beta_j^p,$$

$$(10.5)_p \qquad \tau\gamma_i^p = \Sigma b_{ij}^{\prime p} \beta_j^p + \Sigma c_{ij}^p \gamma_j^p,$$

$$(10.6)_p \qquad \tau\delta_i^p = \Sigma b_{ij}^{\prime\prime p} \beta_j^p + \Sigma c_{ij}^{\prime p} \gamma_j^p + \Sigma d_{ij}^p \delta_j^p,$$

and so

$$\Phi(\tau) = \Sigma(-1)^p \text{ (trace } b^p + \text{trace } c^p + \text{trace } d^p).$$

We now recall the relations

$$F\delta_i^{p+1} = \beta_i^p.$$

Applying F to both sides of $(10.6)_{p+1}$ and recalling that F commutes with τ we have:

$$\tau\beta_i^p = \Sigma d_{ij}^{p+1} \beta_j^p.$$

Hence $d_{ij}^{p+1} = b_{ij}^p$ and trace d^{p+1} = trace b^p. Since there are no matrices d^o, b^n we have the basic relation

$$(10.7) \qquad \Phi(\tau) = \Sigma(-1)^p \text{ trace } c^p.$$

Thus $\Phi(\tau)$ depends solely upon the matrices of transformation of the rational cycles. Notice that since the c_{ij}^p in $(10.5)_p$ are integers, the expression (10.7) for $\Phi(\tau)$ shows that it is an integer.

Since the γ_i^p are, by definition, independent with respect to bounding, their homology classes $\{\Gamma_i^p\}$ form a base for the p-th rational homology group $\mathfrak{H}^p(K)$. The simultaneous homomorphism τ^* induced by τ in these groups is given by

$$(10.8) \qquad \tau^*\Gamma_i^p = \Sigma c_{ij}^p \Gamma_j^p.$$

Thus in (10.7) the c^p are the matrices of transformation of the bases for the homology classes of the rational cycles. To sum up we may state:

(10.9) *Let τ be a chain-mapping of a complex K into itself. There exists an integer $\Phi(\tau)$, depending solely upon the simultaneous homomorphism induced by τ in the rational homology groups and with the property that if $\Phi(\tau) \neq 0$ then τ has fixed elements.*

Everything that has been said holds for the cycles mod π. The same relation (10.5) is obtained with c^p replaced by c_π^p, the analogous matrix for the cycles mod π, and with Φ replaced by $\Phi_\pi = \Phi \bmod \pi$.

11. We pass now to our actual problem, the question of the fixed points in mappings of polyhedra. That is to say if f maps the polyhedron $|K|$ into itself, then x is a *fixed point* of f whenever $fx = x$. We shall find here a perfect analogue to the property (10.7) for chain-mappings.

Let $\{\Gamma_i^p\}$ be a base for the rational homology classes. The simultaneous homomorphism in the rational homology groups gives rise as before to relations

(11.1) $$f\Gamma_i^p = \Sigma c_{ij}^p \Gamma_j^p, \qquad c^p = \|c_{ij}^p\|.$$

The same argument as used before shows that

(11.2) $$\Phi(f) = \Sigma(-1)^p \text{ trace } c^p$$

is independent of the bases. We will now prove:

(11.3) **Fixed Point Theorem.** *The function* $\Phi(f)$ *of the mapping* f *of* $|K|$ *into itself is an integer-valued function with the property that if* $\Phi(f) \neq 0$ *then* f *has a fixed point.*

We shall assume that f has no fixed point and show that this leads to a contradiction.

Under the hypothesis and since $|K|$ is a compactum, $\epsilon = \inf d(x, fx) > 0$. Replacing K if need be by a derived we may assume mesh $K < \epsilon/3$ By (IV, 11), there may then be chosen a subdivision K_1 of K such that there is a barycentric mapping $g : K_1 \to K$ where $g \backsim f$ and fx, gx may be joined by a segment in the closure of a simplex of K. Thus $d(fx, gx) < \epsilon/3$. Hence if σ is any simplex of K the closures $\bar{\sigma}$ and $g\bar{\sigma}$ are disjoint. For if $gx \in \sigma$ then $d(x,fx) < d(x,gx) + d(gx,fx) < 2\epsilon/3 < \epsilon$ which is ruled out.

Let θ be the chain-mapping $K_1 \to K$ induced by g, and let d be chain-subdivision $K \to K_1$. By the remark just made if $\sigma \in K$ and ζ is a simplex of K_1 contained in σ, then $\theta\zeta$ is a simplex σ_1, of K whose closure $\bar{\sigma}_1$ does not meet $\bar{\sigma}$.

Consider now the operation $\theta d : K \to K$. If $\sigma \in K$ then $d\sigma$ is a chain of K_1 whose elements are all contained in σ. Hence $\theta d\sigma$ is a chain of K of which no element meets σ. Therefore θd has no fixed elements. But d induces an isomorphism of the homology groups of K with those of K_1. If we identify the homology classes of K, K_1 with those of the polyhedron $|K|$ itself, then d induces the identity in the homology groups. Hence θd induces the same simultaneous homomorphism as g, and therefore as f, since $f \backsim g$, and it is the one which is represented by (11.1) as regards the rational homology classes. Therefore $\Phi(\theta d) = \Phi(f) = 0$ since θd has no fixed elements. Since this contradicts the hypothesis that $\Phi(f) \neq 0$, our theorem is proved.

12. Applications. *The Brouwer fixed point theorem.* If $K = Cl\,\sigma^n$, hence $|K| = \bar{\sigma}^n$, K is zero-cyclic and if A represents the homology class of a vertex then $fA = A$. Hence $c^0 = \|1\|$, $c^p = 0$ for $p > 0$ and $\Phi(f) = 1$. Therefore f always has a fixed point. This is Brouwer's theorem.

K is a tree. Then the polyhedron $|K|$ is contractible and hence it is zero-cyclic. Hence, as above, $\Phi(f) = 1$ and f always has a fixed point.

Projective plane. The rational homology groups being those of a point we have again $\Phi(f) = 1$. Hence every mapping of a projective plane into itself always has a fixed point.

Spheres. Let $S^n = |\mathfrak{B}\sigma^{n+1}|$, $n > 0$. It is known that S^n is cyclic in the dimensions $0,n$. If Γ^n is its basic integral n-th homology class, and A the class of a vertex, the only significant relations are

$$fA = A, \qquad f\Gamma^n = d\Gamma^n,$$

where d is the degree of f. Hence

$$\Phi(f) = 1 + (-1)^n d.$$

Therefore

(12.1) *Every sense preserving [reversing] mapping of an even [odd] dimensional sphere S^n into itself has a fixed point. Every mapping of S^n into itself whose degree is not ± 1 has a fixed point.*

The application to circumferences and ordinary spheres is immediate. Thus a rotation of a circumference followed by a symmetry about a diameter is an operation with a fixed point. A rotation ρ of S^2 has a fixed point, likewise any mapping $f \backsimeq \rho$.

Surface of genus p. Let Φ_p be the surface, C_i,D_i its basic one-cycles, A a vertex, γ^2 the basic two-cycle (sum of the two-simplexes concordantly oriented). If f maps Φ_p into itself with degree d then

$$fA \sim A, f\gamma^2 \sim d\gamma^2,$$

$$fC^i \sim \Sigma a_{ij}C_j + \Sigma b_{ij}\, D_j,$$

$$fD^i \sim \Sigma c_{ij}C_j + \Sigma d_{ij}\, D_j.$$

Hence

$$\Phi(f) = 1 + d - \Sigma(a_{ii} + d_{ii})$$

and if $\Phi(f) \neq 0$ there is a fixed point.

§4. The Fundamental Group.

13. The study of the multiple determinations of definite integrals between fixed limits leads us to homology, which from the standpoint of comparison of the paths is weaker than homotopy. One may expect therefore that by bringing in homotopy in full strength further noteworthy topological results may be obtained. This is essentially what Poincaré's fundamental group aims to accomplish. This group has already been briefly discussed in the *Survey*, and as stated there, it is the focal point of all the major difficulties of topology. Since the group is generally noncommutative it is written multiplicatively.

The elements and operations of the fundamental group are defined in terms of *paths*. A path α in the arc-wise connected space $\mathfrak{R}$ is simply a mapping of a directed (= oriented) closed arc ab into $\mathfrak{R}$. If E is a subset of $\mathfrak{R}$ and $\alpha(ab) \subset E$ the path is said to be in E. If ab is inverted (= reoriented) to ba the path is said to be described in the opposite direction, denoted by α^{-1}, and also called the *inverse* of α. The point $A = \alpha a$ is the *initial point* and $B = \alpha b$ the *terminal point* of the path; A and B are also referred to as the *end-points* of the path. If A and B coincide the path is *closed*. Notice that a directed arc AB becomes a path under a mapping α equal to the identity.

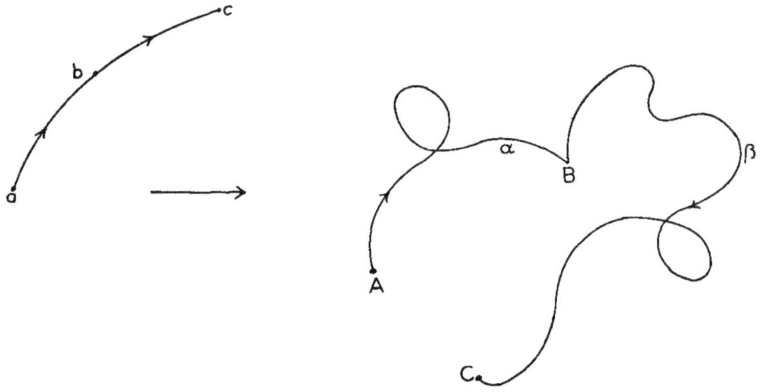

Figure 62

If f is a homeomorphism $ab \to a'b'$ such that $fa = a'$, $fb = b'$ (sense-preserving homeomorphism) then αf is also a path in $\mathfrak{R}$ and it is considered as identical with α. This convention will provide a large measure of freedom in the choice of the initial arc. When the path is closed it is convenient to proceed further as follows: Suppose that J is a directed Jordan curve and $c \in J$. Let g be a mapping $ab \to J$ which is a sense-preserving homeomorphism (see below) of the open interval ab onto $J - c$. If α is a mapping $J \to \mathfrak{R}$ then αg defines a closed path in $\mathfrak{R}$ which we consider as already determined by α regardless of g.

The term *sense-preserving homeomorphism* is utilized here in a somewhat broader sense than before. It is best clarified by means of a parametric representation. Let ab be parametrized by $u : 0 \leq u \leq 1$, so that $u = 0$ corresponds to a and $u = 1$ to b. Then g is to be a function $g(u)$ which is of period 1 and orients J positively.

Let the paths α,β be such that the terminal point B of α is the initial point of β. We may then take an arc abc such that α maps ab and β maps bc into $\mathfrak{R}$. Under the circumstances there is a mapping of the arc abc into $\mathfrak{R}$ agreeing with α on ab and with β on bc and the resulting path is written $\alpha\beta$ and called the *product* of α and β. Intuitively $\alpha\beta$ consists of α followed by β. Its initial point A is the initial point of α and its terminal point C is the terminal point of β. It is readily seen that the product $\alpha\beta$ is unique whenever it is defined.

The extension to a product of several paths $\alpha,\beta, \cdots,$ is immediate and whenever such a product is defined it is associative.

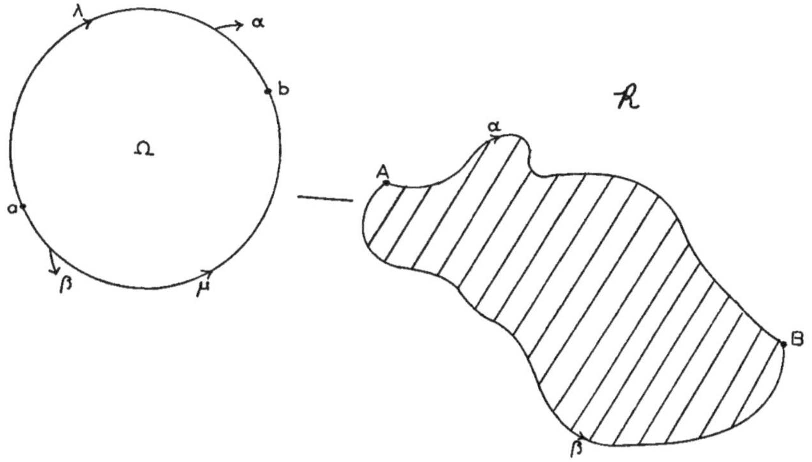

Figure 63

A succession of arcs AB, $BC, \cdots, EF$ gives rise to a "product" path, when each arc $AB, \cdots,$ is considered as a directed arc. The product path $\alpha = (AB) \cdot (BC) \cdot \cdots \cdot (EF)$ is usually and conveniently written $AB \cdots F$. This is the manner in which the paths frequently arise in the applications. If $AB, \cdots$ are one-simplexes of a complex K we shall also refer to α as a *subpath* of K.

Two paths α,β are said to be *homotopic*, written as usual $\alpha \backsim \beta$, whenever they have the same initial and terminal points A,B and may both be represented as mappings of the same arc $\lambda = ab$ under the following conditions: Let l denote the segment $0 \leq u \leq 1$. There is a mapping $\Phi : l \times \lambda \rightarrow \mathfrak{R}$ such that $\alpha = \Phi|0 \times \lambda$, $\beta = \Phi|1 \times \lambda$, $\Phi(l \times a) = A$, $\Phi(l \times b) = B$. This is readily shown to be equivalent to the following: Let Ω be a closed circular region and C its boundary circumference divided into two arcs λ,μ by two points a,b. Then there is a mapping $\Psi : \Omega \to \mathfrak{R}$ such that $\Psi|\lambda = \alpha$, $\Psi|\mu = \beta$. We shall also say: $\beta\alpha^{-1}$ spans the image of a two-cell. If $\alpha \backsim 1$ (= constant mapping), then the situation is the same save that a and b may be taken coincident, there is only one arc λ and α alone spans the image of a two-cell. There is however no compulsion to proceed in this way: we may still assume $a \neq b$ and merely map μ into the point A (coincident end-points of α). As an application let us prove

(13.1) *If α,β are closed paths from A to A (i.e. with both end-points in A) and $\alpha \backsim 1$, $\beta \backsim 1$ then $\alpha\beta$ is likewise a closed path from A to A such that $\alpha\beta \backsim 1$.*

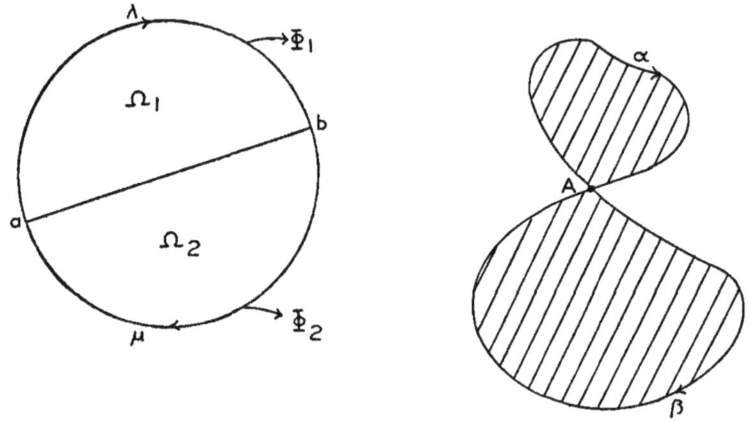

Figure 64

Under the circumstances Ω may be divided into two closed regions Ω_1,Ω_2 by a diameter ab (Fig. 64) and there are mappings $\Phi_i : \Omega_i \to \mathfrak{R}$ such that $\Phi_1|\lambda = \alpha$, $\Phi_2|\mu = \beta$, $\Phi_i(ab) = A$. Hence there is a mapping Φ of $\Omega = \Omega_1 \cup \Omega_2$ into $\mathfrak{R}$ which agrees with Φ_i on Ω_i, and since Φ maps the arc λ followed by μ into $\mathfrak{R}$, we have $\alpha\beta \backsim 1$.

. The natural convention is made that $\alpha 1$ and 1α are identical with α whenever they are defined.

It may be observed that $\alpha\alpha^{-1}$ and $\alpha^{-1}\alpha$ are closed paths. We shall show that each $\backsim 1$. It is manifestly sufficient to consider $\alpha\alpha^{-1}$.

Let in fact abc be an isosceles triangle with $ab = ac$. We may assume that α is a mapping $ba \to \mathfrak{R}$ and that α^{-1} maps $ac \to \mathfrak{R}$ so that if dd' in Fig. 65 is parallel to bc then $\alpha^{-1}d' = \alpha d$. Then $\alpha\alpha^{-1}$ is the mapping of the arc $\lambda = bac$ which agrees with α on ba and with α^{-1} on ac. Let now Φ be a mapping of the triangle abc into $\mathfrak{R}$ such that $\Phi(dd') = \alpha d$, hence $\Phi a = \Phi bc = B$. If l is the segment $0 \leq u \leq 1$, define a mapping

$\Psi : l \times \lambda \to \mathfrak{R}$, as follows: $\Psi(0 \times d) = d$, $\Psi(0 \times d') = d'$, $\Psi(l \times d) = de$, $\Psi(l \times d') = d'e'$. Then $\Phi\Psi$ is a mapping $l \times \lambda \to \mathfrak{R}$ which defines a homotopy of $\alpha\alpha^{-1}$ with the mapping $bac \to B$. Hence $\alpha\alpha^{-1} \backsim 1$ and similarly $\alpha^{-1}\alpha \backsim 1$.

14. It is clear that the paths and their operations obey the following rules:

(a) The product is not always defined.

(b) Whenever a product of several factors has a meaning it is associative.

(c) There is a unit 1 such that whenever $\alpha 1$ and 1α are defined both are equal to α.

(d) There is an inverse α^{-1} and $\alpha\alpha^{-1}$, $\alpha^{-1}\alpha$ are both always defined and equal to unity.

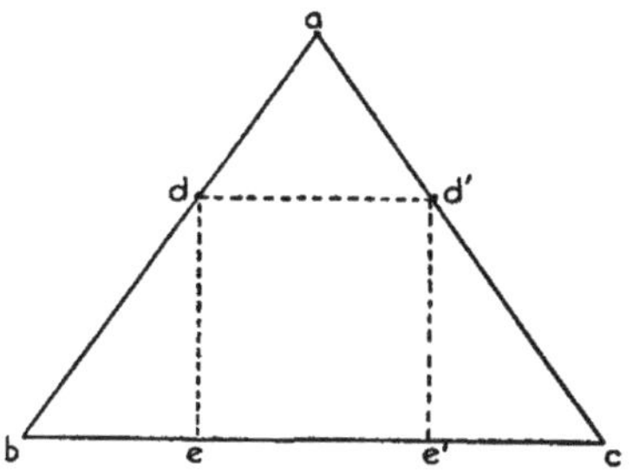

Figure 65

It follows from the definition of the product that the only collections of paths within which the product is uniquely defined and remains in the collection are the set of all closed paths from any given fixed point A, called *base point*, back to A. In view of properties (b), (c), (d) such a collection forms a group G_A. Owing to the convention in (13) the elements $\backsim 1$ form a subgroup G'_A of G_A. If $\alpha \in G'_A$ and $\beta \in G_A$ then $\beta\alpha\beta^{-1} \backsim \beta\beta^{-1} \backsim 1$. Hence $\beta\alpha\beta^{-1} \in G'_A$, or G'_A is an invariant subgroup of G_A. We may therefore introduce the factor-group $H_A = G_A/G'_A$.

Let us show that if A,B are any two points of $\mathfrak{R}$ then $H_A \cong H_B$. In fact since $\mathfrak{R}$ is arc-wise connected we may draw an arc λ from A to B and view it as a path. Hence if $\alpha_1 \in G_B$ then $\alpha = \lambda\alpha_1\lambda^{-1} \in G_A$ and conversely if $\alpha \in G_A$ then $\alpha_1 = \lambda^{-1}\alpha\lambda \in G_B$. Thus $G_A = \lambda G_B \lambda^{-1} \cong G_B$. Moreover if $\alpha_1 \backsim 1$ then $\lambda\alpha_1\lambda^{-1} \backsim 1$. Hence the isomorphism maps G'_A into G'_B. It follows that $H_A \cong H_B$.

Thus when the base point A varies the group H_A is fixed to within an isomorphism. The resulting abstract group is the *fundamental group* of $\mathfrak{R}$. We shall denote it by $\pi_1(\mathfrak{R})$.

To avoid complications we shall usually consider the paths themselves as the elements of $\pi_1(\mathfrak{R})$ and then write $\alpha = 1$ for $\alpha \backsim 1$.

Intuitively the fundamental group may be thought of as generated by the closed paths from any point A to A, with the identity made up of the paths which may be shrunk to A.

15. In many simple cases the fundamental group reduces to the identity. The space $\mathfrak{R}$ is then said to be *simply connected.* Noteworthy instances are found below.

(15.1) *If $\mathfrak{R}$ is contractible to a point, then it is simply connected.*

As a consequence

(15.2) *An open or a closed cell is simply connected.*

We shall now prove:

(15.3) *An n-sphere S^n, $n > 1$, is simply connected.*

In fact let $S^n = |\mathfrak{B}\sigma^{n+1}|$. By the homotopy theorem (IV, 11.1) any path α from a vertex A back to A may be reduced to a path in $|K^1|$, where K^1 is the one-section of $\mathfrak{B}\sigma^{n+1}$. Such a path will be in a closed n-cell containing A and hence $\alpha = 1$. Therefore $\pi_1(S^n) = 1$.

(15.4) *The fundamental group of the one-sphere $S^1 = C$ (a circumference) is infinite cyclic.*

Let C be oriented and let P be the base point in C. Take another oriented circumference D with fixed point Q. The paths are merely the mappings $\alpha : D \to C$ such that $\alpha Q = P$. Their comparison must now be as to homotopies $\Phi : l \times D \to C$ such that $\Phi(l \times Q) = P$. It is an elementary matter to show, as in (IV, 13), that: (a) each path α is uniquely defined as an element of $\pi_1(C)$ by its degree $\rho(\alpha)$; (b) $\rho(\alpha\beta^{-1}) = \rho(\alpha) - \rho(\beta)$. Hence $\alpha \to \rho(\alpha)$ defines an isomorphism of $\pi_1(C)$ with the additive group of the integers and so $\pi_1(C)$ is infinite cyclic.

16. The most interesting case by far is of course when the space $\mathfrak{R}$ is a connected polyhedron Π of positive dimension. If $K = \{\sigma\}$ is any covering complex we shall often write $\pi_1(K)$ for $\pi_1(\Pi)$ and refer to the group as the fundamental group of K.

Let K^p be the p-section, $p > 0$, of K. Since $R^o(K^p) = R^o(K)$, K^p is also connected and so $\pi_1(K^p)$ is defined. We prove:

(16.1) *If* dim $K > 1$, *the fundamental group of K is isomorphic with the fundamental group of its two-section K^2.*

Take as base point P a vertex of K and let $\sigma^2 = abc$ be an oriented triangle and J its oriented boundary. A path α is to be represented as a mapping $J \to \Pi$ sending a into P. Let $L = Cl\ \sigma^2$ and let L_1 be a subdivision of L so chosen that the resulting subdivision J_1 of J and K are star-related relative to the mapping α. The resulting star-projection reduces α to a subpath α_1 of K and hence to an element of $\pi_1(K^2)$. It is an elementary matter to show that further subdivision of L does not affect α_1. Moreover if $\alpha \rightarrow \alpha_1$ and $\beta \rightarrow \beta_1$ then $\alpha\beta \rightarrow \alpha_1\beta_1$. Suppose $\alpha = 1$. Then α_1 may be extended to a mapping $\bar{\sigma}^2 \rightarrow \Pi$. By choosing a suitable subdivision L_1 the extension may be made barycentric and thus α_1 will map L_1 into a subcomplex M_1 of K^2. Hence if $\alpha = 1$ likewise $\alpha_1 = 1$. Therefore $\alpha \rightarrow \alpha_1$ defines a homomorphism $\theta : \pi_1(K) \rightarrow \pi_1(K^2)$. Since every α_1 is an α, θ is onto and since $\alpha_1 = 1$ implies $\alpha = 1$, θ is univalent. Thus θ is an isomorphism and (16.1) follows.

An incidental result of practical importance implicit in the proof may be formulated thus:

(16.2) *To obtain all the operations of $\pi_1(K)$ it is sufficient to consider the closed subpaths from a given vertex Q. If α is such a closed subpath and*

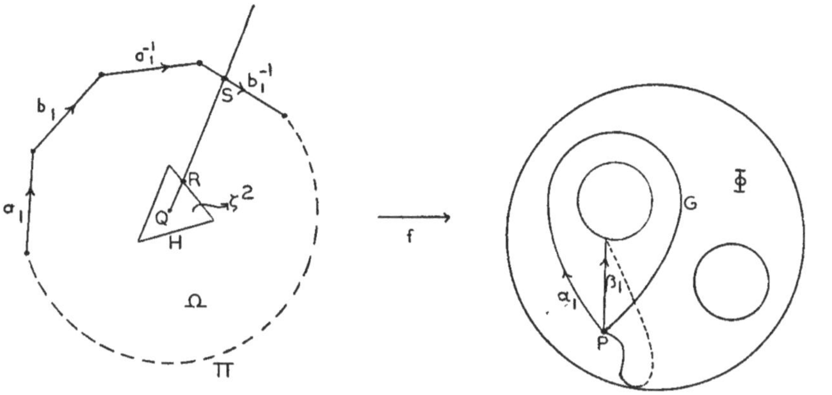

Figure 66

$\alpha = 1$ there exists a subdivision L of a closed triangle and a barycentric mapping $\varphi : L \rightarrow K$ such that if J is the oriented boundary of the triangle with a path origin Q, then $\alpha = \varphi|J$.

17. Application to surfaces. Let first Φ be a closed orientable surface of genus p. Since we have already dealt with spheres we may suppose $p > 0$. The surface has then a normal form (II, 21.1)

$$N = a_1b_1a_1^{-1}b_1^{-1} \cdots b_p^{-1}.$$

It may be constructed by taking a plane convex region Ω bounded by a $4p$ sided polygon Π, whose sides correspond in turn to $a_1, b_1, \cdots$, matching a_i with a_i^{-1} and b_i with b_i^{-1}, then identifying all the vertices. The construction gives rise to a mapping $f : \bar{\Omega} \rightarrow \Phi$ (the same as t of (II, 19)). In the surface Φ there will result $2p$ Jordan curves a_i, β_i intersecting one another in exactly one point P. The path a_i will yield under the mapping f a path (merely α_i oriented) which we also denote by α_i, and similarly for β_i.

Let $\bar{\Omega} = \Omega \cup \Pi$ be triangulated so that the resulting subdivisions on a_i,a_i^{-1} and b_i,b_i^{-1} match. Let K be the resulting complex and K_1 the subcomplex (graph) of K covering Π. The complex K is imaged under f into a complex L covering Φ. The mapping f is a barycentric mapping $K \to L$ topological on Ω and mapping K_1 into a graph G which is a subcomplex of L. Each simplex ζ^s of G other than the vertex P is the image of two simplexes σ'^s, σ''^s of K_1, and f maps ζ^s topologically (affinely) on σ'^s and on σ''^s.

According to (16.2) the paths of $\pi_1(\Phi)$ may all be chosen subpaths of the one-section L^1 of L. Let Q' be a point in $|L| - |L^1|$ and hence in some two-simplex $\zeta_1'^2$ of L. Choose a simplex $\zeta'^2 \subset \zeta_1'^2$ containing Q' and whose closure does not meet G. Under f^{-1} the simplex ζ'^2 is imaged into a simplex $\zeta^2 \subset \Omega$ containing the point $Q = f^{-1}Q'$ and whose closure $\bar{\zeta}^2$ does not meet the polygon Π. Let H be the boundary curve of ζ^2. Any ray QS through Q meets H in a single point R and Π in a single point S. Let φ be the mapping $\bar{\Omega} \to \bar{\Omega}$ defined as follows: $\varphi Q = Q$; $\varphi RS = S$; QR is mapped barycentrically onto QS. Since every point of $\bar{\Omega}$ distinct from its image under φ may be joined to the image by a segment in $\bar{\Omega}$, φ is a deformation. Hence $f \backsimeq f\varphi$ in Φ. Since $f\varphi$ reduces every path from P to P to one in $|G|$, we may assume henceforth that all the operations α of $\pi_1(\Phi)$ are subpaths of G.

Let then α be a subpath of G. If a path contains a subpath $A_iA_jA_i$, the latter may be suppressed as $\backsimeq 1$ in $|G|$. Hence if α contains a subpath of α_i, or of β_i, it contains the whole of α_i or α_i^{-1}, or of β_i or β_i^{-1}. Hence α is of the form

$$\alpha = \gamma_i^{\epsilon_i} \cdots \gamma_j^{\epsilon_j}, \qquad \epsilon_i, \cdots, \epsilon_j = \pm 1,$$

where each γ is an α_i or a β_i. In other words $\pi_1(\Phi)$ is generated by the $2p$ elements α_i,β_i.

There remains to find the nontrivial relations (i.e. not of type $\alpha\alpha^{-1}$) existing between the $2p$ generators. Suppose then $\alpha = 1$. According to (16.2) there exists a subdivision $M_1 = \{\eta\}$ of $M = Cl\,\sigma^2$ and a barycentric mapping $\psi : M_1 \to L$ such that if J is the oriented boundary of σ^2 with an origin of paths Q to be imaged into P, then $\alpha = \psi|J$. If we apply $f^{-1}\psi$ to the elements η of M_1 not mapped into G then we obtain images of these elements in $K - K_1$. Since f^{-1} is topological on each closed simplex $\bar{\zeta}$ of L, the mapping may be extended to the closures of the η's. We have thus a barycentric mapping $\omega = f^{-1}\psi$ defined on a certain closed subcomplex of M_1. Let $\eta = abc$ be a triangle of M_1 on which ω has not yet been defined, and suppose that ω has already been defined on ab. Since $\psi(abc) = a'b'c' \subset G$, $a'b'c'$ is degenerate, hence $c' = a'$ or b', say $c' = b'$. Then $\omega(ab) = a''b''$, a segment of K_1, and we define $\omega a = a''$, $\omega b = \omega c = b''$ and extend ω barycentrically to abc. Since we can pass in σ^2 from one η to any other through a chain of adjacent triangles, we may in this manner extend ω to the whole of $\bar{\sigma}^2$.

Under the definition of ω we will have $\psi = f\omega$. Now if $\lambda = \omega | J$ is a closed path in Π, by (15.4) it is equal to a path of the form h^n, where h is obtained by circular permutation from the path

$$k = a_1 b_1 a_1^{-1} b_1^{-1} \cdot \cdot \cdot b_p^{-1},$$

that is to say $h = lkl^{-1}$. Therefore if $\alpha = 1$, it can only be of the form μ^n, where μ is obtained by circular permutation from

$$\nu = \alpha_1 \beta_1 \alpha_1^{-1} \beta_1^{-1} \cdot \cdot \cdot \beta_p^{-1},$$

that is to say $\alpha = \rho \nu^n \rho^{-1}$. On the other hand $k = 1$ in $\bar{\Omega}$, and hence $\nu = 1$ in Φ. Therefore

$$\alpha_1 \beta_1 \alpha_1^{-1} \beta_1^{-1} \cdot \cdot \cdot \beta_p^{-1} = 1, \tag{17.1}$$

and all the relations between the α_i, β_i are deducible from this unique relation. Thus the fundamental group of the orientable surface of genus p is the abstract group generated by $2p$ elements α_i, β_i satisfying the relation (17.1).

Notice that if $p = 1$ the fundamental relation assumes the form $\alpha_1 \beta_1 = \beta_1 \alpha_1$. Thus the fundamental group of the torus is abelian. It is therefore isomorphic with the free additive group generated by two elements.

For $p = 2$ the group already ceases to be abelian and we have here an example of a noncommutative fundamental group.

Φ *is nonorientable.* The argument is essentially the same. If the normal form is $a_1 a_1 \cdot \cdot \cdot a_k a_k$ then the fundamental group is generated by the elements $\alpha_1, \cdot \cdot \cdot, \alpha_k$ satisfying

$$\alpha_1 \alpha_1 \cdot \cdot \cdot \alpha_k \alpha_k = 1.$$

In particular the projective plane has a group generated by a single element of order two.

18. Relation with homology. It will be recalled that the general concept of homology was put forth as a weak sort of homotopy in which oppositely running paths were allowed to cancel. It suggests that by allowing the operations of the fundamental group to commute one might turn it into a homology group. This is essentially what we propose to prove. It will first be necessary to make a preliminary remark on groups.

Let G be a multiplicative group and $a,b \in G$. The elements of the form $aba^{-1}b^{-1}$ include their inverses and generate by their products a subgroup H known as the *commutator* of G. If $c \in G$ then

$$(cac^{-1})(cbc^{-1})(cac^{-1})^{-1}(cbc^{-1})^{-1} = c(aba^{-1}b^{-1})c^{-1} \in H.$$

Hence H is an invariant subgroup. We may therefore form the factor group G/H and it is the group resulting from G when all its elements are allowed to commute. It is to this group that we refer as "G made commutative."

Any element $h \in H$ may be written in the form $h = a^m \cdots a^n \cdots$, where each factor $a, \cdots$ occurs with a sum of exponents equal to zero. As we shall now show, the converse is also true. In fact, suppose h above has the property in question and let every factor a^m be replaced by $a \cdots a$ written m times and similarly for all factors. Let ν be the number of factors present in h when written in this expanded form. We have then $h = ab \cdots ca^{-1} \cdots e$, where a^{-1} does not occur in the sequence $b \cdots c$. Then

$$h = [a(b \cdots c)\, a^{-1}\, (b \cdots c)^{-1}](b \cdots c \cdots e).$$

The bracket is of the form $a\, ba^{-1}\, b^{-1}$ and the second parenthesis is like h but contains only $\nu - 2$ factors. Hence proceeding in this manner h will be exhibited as a product of terms of form $aba^{-1}b^{-1}$ and so $h \in H$.

(18.1) **Theorem.** *For a connected polyhedron* Π *the fundamental group made commutative is isomorphic with the first integral homology group* $\mathfrak{H}^1(\Pi)$.

Take a covering complex K of Π and choose a vertex P of K. Let also henceforth "closed path" refer once for all to subpaths of K which begin and end at P.

Now any closed path $\alpha = PAB \cdots CP$ determines a unique integral one-chain of $K : \gamma^1(\alpha) = PA + AB + \cdots + CP$, and this chain is a cycle since

$$F\gamma^1(\alpha) = A - P + B - A + \cdots + P - C = 0.$$

Such a cycle will be referred to as *elementary*.

If $\beta = PA' \cdots C'P$ is another closed path so is $\alpha\beta = PA \cdots CPA' \cdots C'P$, and hence at once $\gamma^1(\alpha\beta) = \gamma^1(\alpha) + \gamma^1(\beta)$. Suppose $\alpha = 1$. Setting $L = Cl\, \sigma^2$, and denoting by J the oriented boundary of σ^2, there exists a subdivision L_1 of L with a related subdivision J_1 of J and a barycentric mapping $\varphi : L_1 \to K$ such that $\varphi | J_1$ is precisely α. That is to say J_1 as a path from a certain vertex p of σ^2, back to p, may be written $J_1 = pa \cdots cp$, so that $\varphi p = P$, $\varphi a = A, \cdots, \varphi pa = PA$, $\varphi ab = AB, \cdots$. Let φ stand also for the chain-mapping $L_1 \to K$ which it induces. It is a consequence of the remark just made that φ sends the fundamental cycle $\gamma^1(J_1) = pa + \cdots + cp$ of J_1 into $\gamma^1(\alpha)$. Now $\gamma^1(J_1) = F(d\sigma^2) \sim 0$ in L_1, where d is chain-subdivision $L \to L_1$. Since φ is a chain-mapping, $\varphi\gamma^1(J_1) = \gamma^1(\alpha) \sim 0$ in K. We may therefore assert that $\alpha \to \gamma^1(\alpha)$ defines a homomorphism $\theta : \pi^1(K) \to \mathfrak{H}^1(K) \cong \mathfrak{H}^1(\Pi)$. Thus to prove (18.1) we merely need to show that

(18.2) *θ is a homomorphism of $\pi_1(K)$ onto $\mathfrak{H}^1(K)$ whose kernel is the commutator of $\pi_1(K)$.*

For the proof we require

(18.3) *Every integral one-cycle of a complex K is a linear integral combination of elementary cycles.*

Since the one-cycles depend solely upon the one-section we may suppose that K is a graph. Consider an integral one-cycle

$$\gamma^1 = \Sigma g_i \sigma_i^1,$$

and suppose $\sigma_1^1 = AB$ present in γ^1. Since B is not in $F\gamma^1$ there must be a simplex BC, $C \neq A$, present in γ^1, etc. We thus obtain a sequence of simplexes $AB, BC, \cdots, DE, \cdots$ present in γ^1. Since K is finite we must some time return to a vertex already passed. Hence there exists a sequence of type $AB, \cdots, DA$ consisting of elements in γ^1. Hence

$$\gamma'^1 = \gamma^1 - g_1 (AB + \cdots + DA)$$

is a cycle of $K_1 = K - \sigma_1^1$. Since the parenthesis is an elementary cycle, (18.3) for K will follow if we can prove it for K_1, i.e. for a complex with fewer one-simplexes. Ultimately we shall thus reduce the proof of (18.3) to a complex with a single σ^1 when it is trivial. Thus (18.3) is proved.

Proof of (18.2). Let us first show that θ is *onto*. Join P to each vertex $A \neq P$ by an arc $\lambda(A)$ which is a subpath. Formally also set $\lambda(P) = 1$. If $\gamma^1 = AB + \cdots + DA$ is elementary then $\alpha = \lambda(A) \cdot AB \cdots DA \cdot \lambda^{-1}(A)$ is such that $\gamma^1(\alpha) = \gamma^1$. Now if δ^1 is any integral one-cycle we may write

$$\delta^1 = \Sigma g_i \gamma_i^1$$

where γ_i^1 is elementary. Thus there corresponds to γ_i^1 a closed path α_i (from P to P) such that $\gamma_i^1 = \gamma^1(\alpha_i)$. Hence

$$\delta^1 = \Sigma g_i \gamma^1(\alpha_i) = \gamma^1(\Pi \alpha_i^{g_i}).$$

Therefore θ has the asserted onto property. To complete the proof there remains to show that θ has the proper kernel.

Corresponding to any one-simplex $\sigma^1 = AB$ of K define $\lambda(\sigma^1) = \lambda(A) \cdot AB \cdot \lambda^{-1}(B)$ so that $\lambda(\sigma^1)$ is a closed path. If $\sigma^2 = ABC$ then $\lambda(A) \cdot ABCA \cdot \lambda^{-1}(A)$ is a closed path written $\lambda(F\sigma^2)$. In point of fact the preference seemingly assigned here to the vertex A is unimportant. All that matters is the evident relation

$$\gamma^1(\lambda(F\sigma^2)) = F\sigma^2. \tag{18.4}$$

Moreover since $ABCA \backsim 1$ (as the oriented boundary of σ^2) we also have $\lambda(F\sigma^2) = 1$ as an operation of $\pi_1(K)$.

Now if $\alpha = PA \cdots BC \cdots DP$ we may also write

$$\alpha = \lambda(PA) \cdot \cdots \cdot \lambda(BC) \cdot \cdots \cdot \lambda(DP). \tag{18.5}$$

We also have

$$\gamma^1(\lambda(AB)) = AB + \lambda(A) - \lambda(B),$$

where $\lambda(A), \cdots$, are now chains. Thus

$$\gamma^1(\lambda(AB)) = AB - \lambda(F(AB)),$$

and therefore

$$\gamma^1(\alpha) = \Sigma g_i\sigma_i^1 - \lambda(F(\alpha)) = \Sigma g_i\sigma_i^1.$$

Since $\sigma_i^1 = AB$, g_i is the sum of the exponents of $\lambda(\sigma_i^1)$ in the expression (18.5) for α.

Suppose now $\gamma^1(\alpha) \sim 0$ in K, so that

$$\gamma^1(\alpha) = \Sigma k_j F\sigma_j^2.$$

If we introduce

$$\beta = \Pi[\lambda(F\sigma_j^2)]^{k_j}$$

then by (18.4) $\gamma^1(\beta) = \gamma^1(\alpha)$. Hence $\gamma^1(\alpha\beta^{-1}) = 0$. Thus $\alpha\beta^{-1}$ is a product of closed paths $\lambda(\sigma_i^1)$ in which each appears with a sum of exponents equal to zero. Hence $\alpha\beta^{-1} = \alpha$ is in the commutator group of $\pi_1(K)$. Hence θ has the asserted kernel and the proof of the theorem is completed.

19. Universal covering space. Consider the connected polyhedron Π, a covering complex K of Π and a vertex P of K. With each point Q of Π and path λ from P to Q associate a new point (Q,λ) of a new space $\mathfrak{R}$ under the condition that if λ' is a path from P to Q and $\lambda \backsim \lambda'$ then $(Q,\lambda) = (Q,\lambda')$. Thus there is a point of $\mathfrak{R}$ associated with each point of Π and each operation of the fundamental group. To assign a topology to $\mathfrak{R}$ we proceed as follows. Since K is a complex every point Q possesses an arbitrarily small neighborhood N such that every point Q' of Q may be joined to Q by a segment QQ'. To obtain such a neighborhood we may for example choose a high derived $K^{(n)}$ of K and take $N = |St\ \sigma|$ where σ is the simplex of $K^{(n)}$ containing Q. At all events having N let μ be the segment QQ' in N. Let $\mathfrak{N}$ be the set consisting of (Q,λ) and of all points $(Q',\lambda\mu)$ of $\mathfrak{R}(\lambda\mu$ is the path $PQQ')$. The collection $\{\mathfrak{N}\}$ is chosen as a base for the open sets of $\mathfrak{R}$. It is not difficult to verify that this choice of base turns $\mathfrak{R}$ into a topological space. This space is known as the *universal covering space* of the polyhedron.

If $(Q,\lambda) \in \mathfrak{R}$ then Q is called the *projection* of the point (Q,λ) in Π and (Q,λ) is said to be *over* Q. Let ω denote the projection $(Q,\lambda) \rightarrow Q$. It is clear that ω establishes a homeomorphism of $\mathfrak{N}$ with N. Thus every

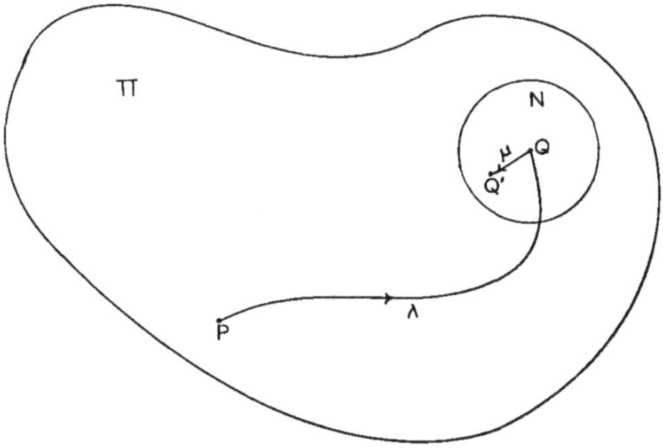

Figure 67

point (Q,λ) of the universal covering space has a neighborhood $\mathfrak{N}$ which is projected topologically on a neighborhood N of Q. A mapping behaving in this manner is often called *locally* topological.

The following properties are readily proved:

*(19.1) *If α is a path in $\mathfrak{R}$ whose end-points are both over the same point Q then $\omega\alpha = \beta$ is a closed path from Q to Q in Π and $\beta = 1$ if and only if α is closed.*

*(19.2) *$\mathfrak{R}$ is simply connected* (i.e. $\pi_1(\mathfrak{R}) = 1$).

*(19.3) *The universal covering space of a Jordan curve is an infinite line.*

*(19.4) *The universal covering space of a projective plane or a two-sphere is a two-sphere.*

Much less obvious is the following noteworthy result whose proof is likewise omitted:

*(19.5) *The universal covering space of an orientable surface of genus* $p > 0$ *is a two-cell.*

§5. THE HOMOTOPY GROUPS.

20. A central problem of topology is the determination of the homotopy classes of mappings. An important step toward its solution would be the solution for mappings of polyhedra into one another. It is not difficult to show that this problem depends upon the possibility of extending a mapping Φ from the bases of a prism $l \times |K|$ to the cells of successive dimensions, and hence from the boundary sphere of a cell $l \times \sigma^p$ to the cell. In short, it would seem desirable from this, and other considerations, to solve first the homotopy problem for spheres.

The preceding general remarks suggest the extension of the fundamental group to mappings of spheres of all dimensions. The required generalization has been carried out and extensively investigated by W. Hurewicz [a] whose work we follow in the present section. For references on the most recent investigations on this topic, see Eilenberg [a], Pontrjagin [a], Steenrod [a], J. H. C. Whitehead [b], George Whitehead [a].

We shall be dealing primarily with the mappings of a "punctured" oriented n-sphere, that is to say, with an oriented n-sphere S^n together with an assigned point Q of S^n, written (S^n,Q). Unless otherwise stated, it is assumed that $n > 1$. The case $n = 0$ offers no interest and $n = 1$ corresponds to the fundamental group already treated. The departures from the latter are such that the case $n > 1$ is altogether dissimilar from the case $n = 1$.

The space $\mathfrak{R}$ into which the mappings are made is assumed arc-wise connected and a pre-assigned point P, called *base point*, is chosen in $\mathfrak{R}$. We shall consider mappings α of the punctured sphere $(S^n,Q) \to \mathfrak{R}$, such that $\alpha Q = P$. If (S'^n,Q') is another such sphere and f is a homeomorphism $S'^n \to S^n$ of degree $+1$ (sense-preserving homeomorphism) such that $fQ' = Q$, then αf is a mapping $S'^n \to \mathfrak{R}$ sending Q' into P. We agree to identify α with αf and this will enable us to deal only with a fixed punctured Euclidean n-sphere in a fixed orientation.

Let α,α' be of the same type as above, l the segment $0 \leq u \leq 1$, and suppose that there exists a mapping $\Phi : l \times S^n \to \mathfrak{R}$ such that $\Phi(l \times Q) = P$ and that Φ agrees with α on $0 \times S^n$ and with α' on $1 \times S^n$. Under the circumstances we shall say that α is *homotopic to* α' *relative to* P, written $\alpha \backsim \alpha'$ rel P. Thus the homotopies here considered are subjected to the restriction that the path of Q is P. It is shown as for ordinary homotopy that homotopy rel P is an equivalence relation and thus gives rise to *homotopy classes* rel P.

If f is as above and $\alpha \backsim \alpha'$ rel P, then likewise $\alpha f \backsim \alpha' f$ rel P and conversely. Hence if we agree to identify α with α' whenever they are homotopic rel P, this will be consistent with the identifications already introduced.

We shall utilize the preceding remark to reduce α to a certain simple normal form. To that effect take any point R of $S^n - Q$ and a closed spherical region Ω of center R which does not contain Q. Define now a mapping f of S^n into itself in the following way: $f(S^n - \Omega) = Q, fR = R$. If S^{n-1} is the boundary sphere of Ω and $M \in \Omega - R$, there is a unique plane containing Q,M,R; it intersects S^n in a circumference C with a

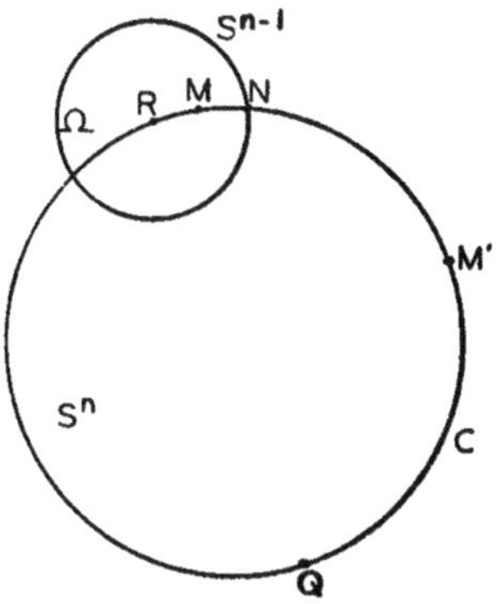

Figure 68

unique arc RMQ. This arc intersects S^{n-1} in a single point N and we define $fM = M'$ as the point which divides the arc RMQ in the same ratio as M divides the arc RMN. If $M \neq fM$, there is always a unique arc MM' on the arc RMQ. Hence f is a deformation and under our conventions $\alpha f = \alpha$. Thus we may replace α by a mapping concentrated on an arbitrarily small spherical region Ω in the sense that $S^n - \Omega$ is mapped into the fixed point P. The concentration might be made on any closed region or set of nonoverlapping closed regions Ω_1 whatsoever not containing Q. All that is necessary to that effect is to concentrate α on some closed spherical region $\Omega \supset \Omega_1$. The closed regions of concentration will be referred to as *islands*.

It is sometimes convenient to take for image of S^n a Euclidean space $\mathfrak{E}^n$ closed at infinity by the point Q. The islands are then merely bounded regions in $\mathfrak{E}^n$ and they may undergo quite arbitrary displacements in $\mathfrak{E}^n$, subjected to the sole condition that the various islands under consideration must never overlap.

21. We shall now introduce in the collection $\mathfrak{C}$ of all the homotopy classes rel P certain relations which will turn $\mathfrak{C}$ into an additive group. The relative homotopy class determined by a mapping α will be denoted temporarily by $[\alpha]$.

Let α_1,α_2 be two mappings of the punctured sphere (S^n,Q) into $\mathfrak{R}$. Concentrate them on two disjoint islands Ω_1,Ω_2 and let α be the mapping which agrees with α_i on Ω_i and sends $S^n - \Omega_1 - \Omega_2$ into the base point P. It is at once seen that the class $[\alpha]$ depends solely upon the classes $[\alpha_1]$ and $[\alpha_2]$ and we write $[\alpha] = [\alpha_1] + [\alpha_2]$.

By displacement in $S^n - Q$ one may interchange Ω_1 with Ω_2 and hence replace $\alpha_1 + \alpha_2$ by $\alpha_2 + \alpha_1 \backsimeq \alpha_1 + \alpha_2$ rel P. Hence $[\alpha_2] + [\alpha_1] = [\alpha_1] + [\alpha_2]$, or class addition is commutative.

The sum of three or more $[\alpha_i]$ is introduced as usual. By concentration of the α_i on the appropriate number of islands associativity of the sum becomes obvious.

The freedom in the construction yields also the following convenient result. If $\alpha_i, i = 1,2$, is concentrated on several islands Ω_{ij}, then $[\alpha_1] + [\alpha_2]$ is the class of the mapping which agrees with α_i on the Ω_{ij} and sends everything else into P.

Let ω denote the constant mapping $S^n \to P$. The class $[\omega]$ is by definition the zero class of the forthcoming group, and is written 0. Denoting by Δ^{n+1} the disk bounded by S^n, the n.a.s.c. for a mapping α to be in the zero-class is that α may be extended to a mapping $\Delta^{n+1} \to R$ (I,15.8).

Let $\mathfrak{E}^n$ be a diametral subspace of the space of S^n containing Q and let φ be the operation of symmetry of S^n relative to $\mathfrak{E}^n$. If α varies in a fixed class of $\mathfrak{C}$, so does $\alpha\varphi$ and $[\alpha\varphi]$ depends solely upon $[\alpha]$. By definition $-[\alpha] = [\alpha\varphi]$. It is not difficult to show that in this statement φ may be replaced by any mapping $\varphi' : S^n \to S^n$ leaving Q fixed and whose degree is -1. However, φ is quite sufficient for our purpose.

(21.1) *Under the definitions of* $+,0,-$, *just given the collection* $\mathfrak{C}$ *of homotopy classes* rel P *is turned into an additive group* $\pi_n(P)$.

Since the addition has already been proved commutative and associative, we merely need to show that:

$$\text{(a)}\quad [\alpha] + 0 = [\alpha]$$

$$\text{(b)}\quad [\alpha] + (-[\alpha]) = 0.$$

Proof of (*a*). Concentrate α on an island Ω_1, and let Ω_2 be an island disjoint from Ω_1. The concentration of ω on Ω_2 is merely ω itself. Hence $\alpha \in [\alpha] + [\omega]$ and this implies (a).

Proof of (*b*). Let H,H' be the two hemispheres into which $\mathfrak{E}^n$ divides S^n. Concentrate α on an island $\Omega \subset H$, and hence $\alpha\varphi$ on the island $\varphi\Omega \subset H'$. Now if $M \in H$ and $M' = \varphi M$, define a mapping $\Phi : \Delta^{n+1} \to \mathfrak{R}$ by the condition that $\Phi(MM') = \alpha M$. Since Φ agrees with α on Ω and with $\alpha\varphi$ on $\varphi\Omega$, $[\alpha] + [\alpha\varphi] = 0$ and this is (b). This completes the proof of (21.1).

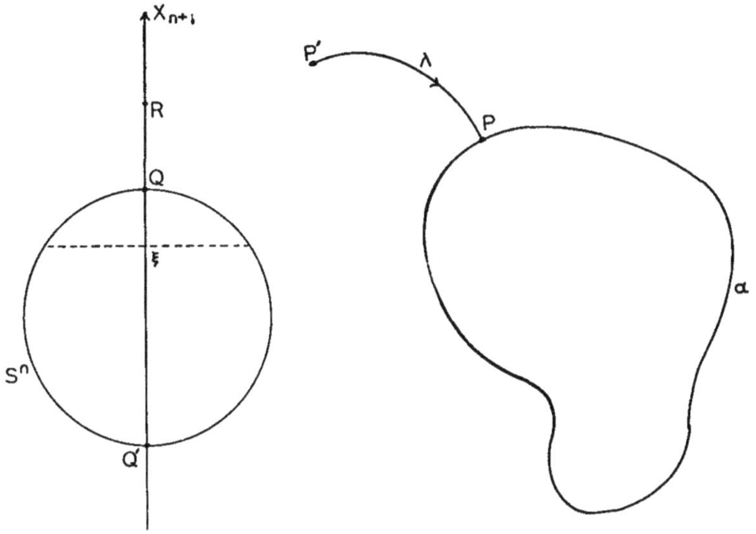

Figure 69

We shall now show that the group $\pi_n(P)$ is independent of P. More precisely:

(21.2) *If P,P' are any two points of $\mathfrak{R}$ then $\pi_n(P) \cong \pi_n(P')$.*

Let S^n be the sphere

$$x_1^2 + \cdots + x_{n+1}^2 = 1,$$

and let Q be the point $(0, \cdots, 0,1)$, R the point $(0, \cdots, 0,2)$ and ξ a point of the vertical axis (axis through Q) between Q and its antipode Q'. Set $\Lambda = \Delta^{n+1} \cup QR$ and define a deformation $g : \Delta^{n+1} \to \Lambda$ as follows: $\xi \to Q$, $Q \to R$, Q' remains fixed; all points on a space $x_{n+1} =$ const., other than Q', move upwards and remain on such a space.

Since the space $\mathfrak{R}$ is arc-wise connected, there is a path λ from P' to P. Let λ be represented by a mapping $\varphi : RQ \to \mathfrak{R}$. Let also ψ denote the mapping $\Lambda \to \mathfrak{R}$ which agrees with φ on RQ and with α on S^n. Then $\beta = \psi g$ is a mapping of the punctured sphere (S^n,Q) into R sending Q into P'. Let the homotopy classes rel P' be denoted by $[\beta]'$. Thus $[\beta]'$ is the class of β.

It is immediately seen that $[\beta]'$ depends solely upon $[\alpha]$ and λ. Let us write $[\beta]' = \lambda[\alpha]$.

Let similarly α_1 give rise to β_1 and let α,α_1 be concentrated on disjoint islands Ω,Ω_1 in the hemisphere H. As a consequence the representative $\alpha\varphi$ of $-[\alpha]$ is concentrated on an island $\Omega' = \varphi\Omega$ in H' and hence disjoint from Ω_1. Now the class $[\alpha_1] - [\alpha]$ is represented by an α^* concentrated on Ω' and Ω_1 and agreeing with α_1 on Ω_1 and with $\alpha\varphi$ on Ω'. Hence $\lambda\{[\alpha_1] - [\alpha]\}$ is represented by a mapping β^* which agrees with β_1 on Ω_1 and with $\beta\varphi$ on Ω'. Therefore

$$\lambda\{[\alpha_1] - [\alpha]\} = [\beta^*]' = [\beta_1]' - [\beta]' = \lambda[\alpha_1] - \lambda[\alpha].$$

Hence the operation λ is a homomorphism $\pi_n(P) \to \pi_n(P')$.

Consider now the effect of certain modifications in λ. Let R' be the midpoint of RQ and let now φ be such that $\varphi|RR' = \lambda$, $\varphi|R'Q = \mu$. Then we would have to replace above λ by $\lambda\mu$. The product $\lambda\mu[\alpha]$ is readily shown to be associative.

Take in particular any mapping β of a class $[\beta]' \in \pi_n(P')$. Then, operating with λ^{-1} and β as previously with λ and α, there results an α such that $\lambda[\alpha] = \lambda\lambda^{-1}[\beta]' = [\beta]'$. Hence λ is a homomorphism of $\pi_n(P)$ onto $\pi_n(P')$.

Returning to λ itself, suppose that $P' = P$, so that λ is a closed path from P to P. It is immediately seen that, if λ varies in its homotopy class rel P, i.e. if it is *fixed* as an operation of the fundamental group, then the class $\lambda[\alpha]$ is likewise fixed.

Consider again the case $P' \neq P$, so that now λ is not closed. Applying λ to $[\alpha]$, then λ^{-1} to the result, we have $\lambda^{-1}\lambda[\alpha] = [\alpha]$ since $\lambda^{-1}\lambda$ is closed and $\backsim 1$ rel P. This shows also that, if $\lambda[\alpha] = 0$, then $[\alpha] = 0$. Hence λ is univalent and therefore it is an isomorphism. This proves (21.2).

The abstract group which has been obtained is written π_n, or $\pi_n(\mathfrak{R})$, and is called the n-th *homotopy group* of $\mathfrak{R}$.

We have assumed throughout $n > 1$. The treatment could be made to cover also the fundamental group $\pi_1(\mathfrak{R})$. Actually whenever $\pi_1(\mathfrak{R})$ occurs implicitly as a homotopy group, it is understood to mean the fundamental group.

Henceforth we abandon the symbol [ ] and merely designate indifferently by α an element of π_n or a representative of the element as a mapping relative to some definite preassigned point P.

Remark. Let $\lambda \in \pi_1(\mathfrak{R})$ and $\alpha \in \pi_n(\mathfrak{R})$. Keeping λ fixed, it has been shown in substance that $\alpha \to \lambda\alpha$ defines an automorphism $\pi_n \to \pi_n$. This is often expressed as: π_1 is a group of operators for π_n.

The following property is an immediate consequence of the definition of the groups:

(21.3) *If the space $\mathfrak{R}$ is contractible (see I, 15) then all its homotopy groups vanish. Hence the homotopy groups of an open or a closed cell, of an Euclidean space or of a tree are all zero.*

In this and similar general statements later "π_1 vanishes" stands for "π_1 reduces to its unit element."

22. (22.1) **Theorem.** *A n.a.s.c. in order that the first n homotopy groups vanish is that every mapping into $\mathfrak{R}$ of a polyhedron Π of dimension $\leq n$ be homotopic to a constant.*

An immediate corollary is:

(22.2) *If Π is connected, n dimensional and its first n homotopy groups vanish, then Π is contractible, and conversely.*

Proof of (22.1). Sufficiency is obvious and so only necessity requires proof. Let P be taken as base point in $\mathfrak{R}$. Let φ be a mapping of the polyhedron $\Pi \to \mathfrak{R}$, where $\dim \Pi \leq n$. Take any complex $K = \{\sigma\}$ covering Π and construct a join $L = K \cup AK \cup A = \{\sigma, A\sigma, A\}$. To prove (22.1) it is sufficient to show that the mapping φ may be extended to $|L|$, with $\varphi A = P$ (I, 15.8). All the vertices of L are already mapped and so one must extend φ to the simplexes of positive dimension. The σ_i^1 of K are already mapped so that we are only concerned with the one-simplexes $A\sigma_i^0$. If $\varphi\sigma_i^0 = P$ define $\varphi A\sigma_i^0 = P$. If $\varphi\sigma_i^0 \neq P$ and since $\mathfrak{R}$ is arc-wise connected we may join the two points by an arc ζ_i^1 then extend φ to $A\sigma_i^0$ by a topological mapping $A\sigma_i^0 \to \zeta_i^1$.

Suppose that all the simplexes of L whose dimension $\leq p$, where $p > 1$, are already mapped. The $\sigma_i^{p+1} \in K$ are already mapped so that we are only concerned with the $(p+1)$-simplexes $A\sigma_i^p$. If $S^p = |\mathfrak{B}A\sigma_i^p|$, then $\varphi|S^p$ is already defined and it sends A into P. Since $\pi_p(\mathfrak{R}) = 0$, this mapping may be extended to $A\sigma_i^p$ and so step by step φ may be extended to the whole of L. This completes the proof of (22.1).

23. The following proposition establishes a noteworthy connection between the homotopy groups and homology. It holds actually for more general spaces than polyhedra but is given here only for the latter as we have not discussed homology for other spaces.

(23.1) **Theorem.** *If the first $(n-1)$ homotopy groups of a connected* *polyhedron* Π *vanish then the n-th* $\pi_n(\Pi)$ *is isomorphic with the integral homology group* $\mathfrak{H}^n(\Pi)$. (*Note that n need not be the dimension of* Π.)

Immediate consequences are:

(23.2) *A n.a.s.c. in order that the integral homology groups* $\mathfrak{H}^p(\Pi)$, $1 \leq p \leq n$, Π *connected, vanish, is that the first n homotopy groups of* Π *vanish.*

(23.3) *The first* $n - 1$ *homotopy groups of the n-sphere* S^n, $n > 1$, *vanish and the n-th homotopy group* $\pi_n(S^n)$, $n \geq 1$, *is cyclic.*

Proof of (23.1). Let $K = \{\sigma\}$ be a complex covering the polyhedron Π and let the base point P be a vertex of K. We shall only be concerned with the operations of $\pi_n = \pi_n(\Pi)$ and they will be exhibited as mappings of a punctured sphere (S^n,Q). As the sphere we shall choose the boundary sphere of a simplex $Q\zeta^n$. Set $L = \mathfrak{B}Q\zeta^n$ and assume L, hence S^n, oriented by the fundamental cycle $\gamma^n = FQ\zeta^n$.

We may represent an α of π_n as a barycentric mapping of a subdivision L_1 of L into K. Let α stand also for all the operations which it induces on cycles, $\cdots$. If d is chain-subdivision $L \to L_1$ then $\gamma_1^n = d\gamma^n$ is the orienting fundamental cycle of L_1 and we will have

$$\alpha\gamma_1^n = \alpha d\gamma^n = \Sigma s_i\sigma_i^n = \delta^n(\alpha),$$

where $\delta^n(\alpha)$ is a cycle of K. Let $\alpha' \in \pi_n$ give rise in similar manner to $\delta^n(\alpha')$ with coefficients s_i'. By paraphrasing the treatment of (IV, 19.6) for the mappings of spheres on spheres, it may be shown that α,α' may be concentrated on islands which are simplexes of a certain L_1, in the sense that only the islands go into n-simplexes σ_i^n, so that: (a) the simplexes of the two concentrations are disjoint; (b) if $s_i \neq 0$ the new α maps exactly $|s_i|$ simplexes of its concentration on σ_i^n and with the same sign as s_i; (c) similarly with α', its concentration and the s_i'. Since manifestly $\delta^n(-\alpha) = -\delta^n(\alpha)$, we conclude from this that the situation may be so disposed of that

$$\delta^n(\alpha - \alpha') = \delta^n(\alpha) - \delta^n(\alpha').$$

If we pass to the homology classes Γ, Δ, which are in fact classes of S^n and Π, we will have

$$\alpha\Gamma^n = \Delta^n(\alpha), \qquad \Delta^n(\alpha - \alpha') = \Delta^n(\alpha) - \Delta^n(\alpha').$$

Hence $\alpha \to \Delta^n(\alpha)$ defines a homomorphism θ of π_n into $\mathfrak{H}^n$. Our task will be fulfilled if we can show that

(23.4) θ *is an isomorphism of* π_n *with* $\mathfrak{H}^n$.

The proof will consist in showing that θ is univalent and has the onto property.

24. We begin with univalence. We suppose $\delta^n(\alpha) \sim 0$, or

$$\delta^n(\alpha) = \Sigma\mu_i F\sigma_i^{n+1}, \tag{24.1}$$

and we must show that as a consequence $\alpha = 0$.

Let S_i^n denote the boundary sphere of σ_i^{n+1} oriented by the fundamental cycle $\gamma_i^n = F\sigma_i^{n+1}$. For convenience identify temporarily the punctured sphere (S^n,Q) with S_i^n so that Q becomes a vertex of σ_i^{n+1}, and more particularly P if P is a vertex of σ_i^{n+1}. Let T_i consist of S_i^n with an arc λ_i from Q to P which is a subpath of K if P is not in S_i^n, and let $T_i = S_i^n$ if $P \in S_i^n$. Let S_i^n undergo a deformation $\varphi : S_i^n \to T_i$ such as in the proof of (21.2) giving rise to an operation $\alpha(\sigma_i^{n+1})$ of π_n. The operation φ may be taken as a barycentric mapping of a subdivision of S_i^n into the subcomplex of K contained in T_i. Since the n-dimensional part of T_i is S_i^n, the theory of the degree may be applied to it without any modification. Since φ is a deformation its degree is one, hence $\varphi\gamma_i^n = \gamma_i^n$. Thus $\delta^n(\alpha(\sigma_i^{n+1})) = \gamma_i^n$.

Since $S_i^n \backsim 1$ on $\sigma_i^{n+1} \cup T_i$, $\alpha(\sigma_i^{n+1}) = 0$. Returning then to the situation of (24.1), if

$$\alpha' = \Sigma\mu_i\alpha(\sigma_i^{n+1})$$

then $\alpha' = 0$. On the other hand also

$$\delta^n(\alpha') = \Sigma\mu_i\delta^n(\alpha(\sigma_i^{n+1})) = \Sigma\mu_i F\sigma_i^{n+1} = \delta^n(\alpha).$$

Therefore $(\alpha - \alpha') = \alpha$ and at the same time $\delta^n(\alpha - \alpha') = 0$. Consequently, whenever $\delta^n(\alpha) \sim 0$, α may be replaced (as a representative of an element of π_n) by an element $\alpha_1 = \alpha$ such that $\delta^n(\alpha_1) = 0$. We may assume then that actually $\delta^n(\alpha) = 0$.

Under the circumstances α may be reduced to a mapping of the punctured sphere into $|K^{n-1}|$, where K^{n-1} is the $(n-1)$-section of K. Since the first $n-1$ homotopy groups of Π are zero, referring to (22.1) and its proof there is a deformation ψ of $|K^{n-1}|$ into a constant, which may be assumed to be the base point P itself and under which P is its own path. Hence $1\alpha \backsim \psi\alpha$ or α is also homotopic to the constant P. That is to say, $\alpha = 0$. Thus the homomorphism θ of (23) is univalent.

25. We shall now show that θ has the onto property. Let $M = K \cup QK^{n-1} \cup Q$. In view of (22.1) and our hypotheses $|K^{n-1}|$ is deformable into a point of $|K|$, which implies that the identity on $|K^{n-1}|$ may be extended to a mapping $|K^{n-1} \cup QK^{n-1} \cup Q| \to \Pi$. From this follows that there is a mapping $\varphi : |M| \to \Pi$ which reduces to the identity on $|K^n|$ such that $\varphi Q = P$. One may then approximate φ by a mapping, still called φ, which is barycentric $M_1 \to M$, where M_1 is a subdivision of M, and where again $\varphi Q = P$. Let K_1 be the subdivision induced in K by M_1 and let d be chain subdivision $M \to M_1$. The approximation process is such that on K_1 the operation φ is an inverse of d and so $\varphi d = 1$ as regards operations on chains of K.

Consider now the sphere $S_i^n = \sigma_i^n \cup Q\mathfrak{B}\sigma_i^n \cup Q$, and let it be punctured at Q. If $\gamma_i^n = \sigma_i^n - QF\sigma_i^n$, then $F\gamma_i^n = 0$, or γ_i^n is an n-cycle. Since γ_i^n contains σ_i^n with coefficient $+1$, it is a fundamental cycle for the sphere S_i^n. Let it serve to orient the sphere. We shall require in a moment the boundary relations

$$F\sigma_i^n = \Sigma \eta_{ij}\sigma_j^{n-1}.$$

Now under φ the mapping of S_i^n gives rise to an operation of the homotopy group π_n which we denote by α_i. The homology class $\Delta^n(\alpha_i)$ is the class of the cycle $\varphi d\gamma_i^n$. We have just shown that $\varphi d\sigma_i^n = \sigma_i^n$. Hence $\Delta^n(\alpha_i)$ is the class of the cycle

$$\delta_i^n = \sigma_i^n - \Sigma \eta_{ij}\varphi dQ\sigma_j^{n-1}.$$

Consider now any integral n-cycle of K:

$$\delta^n = \Sigma \mu_i \sigma_i^n.$$

Since δ^n is a cycle, $F\delta^n = 0$, and so

$$\Sigma \mu_i \eta_{ij} = 0.$$

Consider now the operation

$$\alpha = \Sigma \mu_i \alpha_i$$

of the group π_n. Since θ is a homomorphism,

$$\Delta^n(\alpha) = \Sigma \mu_i \Delta^n(\alpha_i).$$

Hence $\Delta^n(\alpha)$ is the class of the cycle

$$\Sigma \mu_i \delta_i^n = \Sigma \mu_i \sigma_i^n - \Sigma \mu_i \eta_{ij}\varphi dQ\sigma_j^{n-1} = \delta^n.$$

Therefore θ has the onto property.

Since θ is both univalent and onto it is an isomorphism, thus proving (23.4) and hence also theorem (23.1).

26. (26.1) Theorem. *The homotopy groups $\pi_n(\Pi)$, $n > 1$, of a connected polyhedron Π are isomorphic with those $\pi_n(\mathfrak{R})$ of its universal covering space $\mathfrak{R}$.*

Combined with (19.3), (19.5) and (21.3) this yields:

(26.2) Corollary. *The homotopy groups π_n, $n \geq 2$, of a circumference, or of a closed surface other than a sphere, are all zero.*

Proof of (26.1). Let the various notations be as in (23). In addition, choose as base point in $\mathfrak{R}$ a point P^* which is over the base point P in Π. Denote also by ω the projection $\mathfrak{R} \to \Pi$. Let α be a mapping of the punctured sphere $(S^n,Q) \to \Pi$ such that $\alpha Q = P$. We propose to associate with α a definite mapping $\alpha^* : (S^n,Q) \to \mathfrak{R}$ such that $\omega\alpha^* = \alpha$, $\alpha^*Q = P^*$, and to show that $\alpha \to \alpha^*$ gives rise to an isomorphism of $\pi_n(\Pi)$ with $\pi_n(\mathfrak{R})$. Certain preliminary considerations are, however, necessary.

Let $\mathfrak{A}$ be a connected and arc-wise connected compactum with group $\pi_1(\mathfrak{A}) = 1$. Take any point $Q \in \mathfrak{A}$ and let f be a mapping $\mathfrak{A} \to \Pi$ such that $fQ = P$. Given any point P^* of $\mathfrak{R}$ over P we propose to show that:

(26.3) *There is a mapping* $f^* : \mathfrak{A} \to \mathfrak{R}$, *such that* $f^*Q = P^*$ *and* $\omega f^* = f$. *This mapping* f^* *is unique and sends into* P^* *all points of* $\mathfrak{A}$ *which* f *sends into* P.

Let $Y \in \Pi$ and let Y^*,Y^{**} be points of the universal covering space $\mathfrak{R}$ over Y. One may draw in $\mathfrak{R}$ a path λ^* from Y^* to Y^{**} and $\lambda = \omega\lambda^*$ is an element of $\pi_1(\Pi)$ whose assignment suffices to characterize Y^{**}. Owing to this it is convenient to denote Y^{**} by λY. The distinct points over Y correspond to the distinct elements of the fundamental group $\pi_1(\Pi)$ and Y^* corresponds to unity.

We also recall that Y,Y^* have neighborhoods V,V^* such that ωV maps V^* topologically on V (19). Let λV^* stand for the set of points λZ^*, $Z^* \in V^*$. Then referring to the definition of these neighborhoods, it will be seen that λV^* is related to λY^* and V, like V^* to Y^* and V.

Since Π is compact, it has a finite covering $\{V_1, \cdots, V_r\}$ by sets such as V. The analogues of V^* are written V_i. Let $f^{-1}V_i = U_i$. Since f is continuous, U_i is an open set of $\mathfrak{A}$ and $\{U_i\}$ is a finite open covering of $\mathfrak{A}$.

Take now any $X \in \mathfrak{A} - Q$. Since $\mathfrak{A}$ is arc-wise connected, there is an arc γ from Q to X. Let ϵ be the Lebesgue number of $\{U_i\}$ and subdivide γ by points $Q = X_1, X_2, \cdots, X_s = X$ such that the closed arcs $\gamma_i = X_iX_{i+1}$ are of diameter $< \epsilon$. Each will then be in a set U_i. At the cost of some repetition one may number the U_i so that $\gamma_i \subset U_i$.

Since $P \subset V_1$ the notations may be so chosen that $\lambda P^* \in \lambda V_1^*$ and in particular $P^* \in V_1^*$. Under the prescribed conditions $f_1^* = f^*|U_1$ is uniquely determined as the mapping $\omega_1^{-1}f$. Under f_1^* the arc γ_1 gives rise to a path δ_1 from $Y_1^* = P^*$ to the point $Y_2^* = f_1^*X_2$. The operation is now repeated with $X_2, \cdots$, in place of $Q = X_1, \cdots$, etc., and yields finally in unique manner a path δ from P^* to a point Y^*, which is defined as f^*X. It is a consequence of the construction that f^* has all the required properties provided that it is single-valued. We shall show that it possesses this property also.

Assuming f^* not single-valued, there exists a second path γ' from Q to X in $\mathfrak{A}$ such that the above treatment leads to a path δ' from P^* to a point Y^{**} still over Y but distinct from Y^*. As a consequence $\eta = \gamma^{-1}\gamma'$ is a closed path from X to X in $\mathfrak{A}$ giving rise under the same circumstances as above to a path ζ^* from Y^* to Y^{**} in $\mathfrak{R}$ such that $\zeta = \omega\zeta^* = f\eta$ is a closed path from Y to Y in Π. It follows that $\zeta \neq 1$ as an operation of $\pi_1(\Pi)$. On the other hand, since $\pi_1(\mathfrak{A}) = 1$, $\eta \backsim 1$ rel X in $\mathfrak{A}$, and hence $\zeta \backsim 1$ rel Y in Π. This contradiction shows that f^* is single-valued and proves (26.3).

Returning now to our theorem take in (26.3) $\mathfrak{A} = S^n$, $f = \alpha$. Then f^* becomes an $\alpha^* \in \pi_n(\mathfrak{R})$ such that $\omega\alpha^* = \alpha$.

Suppose now $\alpha \backsim \beta$ rel P. Let them give rise as above to α^*, β^*. By hypothesis there is a mapping $\Phi : l \times S^n \to \Pi$, $l : 0 \leq u \leq 1$, where $\Phi(l \times Q) = P$, and Φ agrees with α on $0 \times S^n$ and with β on $1 \times S^n$. Taking now in (26.3) $\mathfrak{A} = l \times S^n$, $f = \Phi$, there results a mapping $\Phi^* : l \times S^n \to \mathfrak{R}$ such that $\omega\Phi^* = \Phi$, $\Phi^*(l \times Q) = P^*$. Let α'^* be the mapping $S^n \to \mathfrak{R}$ which agrees with Φ^* on $0 \times S^n$. Thus $\omega\alpha'^* = \alpha$ and $\alpha'^*Q = P^*$. Hence $\alpha'^* = \alpha^*$. In other words Φ^* agrees with α^* on $0 \times S^n$, and it agrees similarly with β^* on $1 \times S^n$. We have then $\alpha^* \backsim \beta^*$ rel P^*. Hence the homotopy class of α^* rel P^* depends solely upon the homotopy class of α rel P, and so $\alpha \to \alpha^*$ determines a transformation $\theta : \pi_n(\Pi) \to \pi_n(\mathfrak{R})$. Having recourse to the concentration on islands it is shown immediately that $\theta(\alpha - \beta) = \theta\alpha - \theta\beta$. Hence θ is a homomorphism. From $\omega\alpha^* = \alpha$ follows $\alpha = 0$ if $\alpha^* = 0$, so that θ is univalent. Finally, given $\alpha^* \in \pi_n(\mathfrak{R})$, with base point P^*, $\omega\alpha^* = \alpha \in \pi_n(\Pi)$ with base point P. Therefore every α^* is an $\omega\alpha$. Hence θ has the onto property and so it is an isomorphism. This completes the proof of the theorem.

PROBLEMS

1. *The Alexandroff mapping theorem.* Given a finite collection $\mathfrak{A} = \{A_i\}$, whenever the intersection $A_i \cap \cdots \cap A_j$ is nonvacuous introduce a simplex $\sigma = A_i \cdots A_j$, with $A_i, \cdots, A_j$ as vertices. Then $K = \{\sigma\}$ is a (nongeometric) complex known as the *nerve* of $\mathfrak{A}$.

Given a finite open ϵ covering $\mathfrak{U}$ of a compactum $\mathfrak{R}$, let the nerve of $\mathfrak{U}$ be realized as a geometric complex $\mathfrak{K}$. Show that there is a 3ϵ mapping of $\mathfrak{R}$ into the polyhedron $|\mathfrak{K}|$ (the inverse image of each point of the image is of diameter $< 3\epsilon$).

Alexandroff's complete theorem is as follows: if dim $\mathfrak{R} = n$ (in the sense of Menger-Urysohn) $\mathfrak{R}$ can be ϵ mapped into a polyhedron of dimension n, but not, for arbitrarily small ϵ, into one of lower dimension.

2. Let $\bar{a}$ denote the complex conjugate of a number a. Let Σ^2 be a complex cartesian plane referred to coordinates x,y with distance $[(x - x')(\bar{x} - \bar{x}') + (y - y')(\bar{y} - \bar{y}')]^{1/2}$. Show that Σ^2 is a four-cell.

3. In the same notations as in problem 2 let S^3 denote the three-sphere $x\bar{x} + y\bar{y} = 1$ of Σ^2 and let S^2 be a two-sphere parameterized by the complex variable z. The complex line $y = zx$ intersects S^3 in a circumference $C(z)$ and $C(z) \to z$ defines an essential mapping $S^3 \to S^2$ (H. Hopf). Can the complex lines be replaced by other curves and does this yield new essential mappings?

Remark. Hopf has determined all the essential mappings $S^3 \to S^2$ and shown that they are completely characterized by the value of a certain integer γ. In the mapping defined above $\gamma = 1$ and for inessential mappings $\gamma = 0$.

4. Let $\mathfrak{P}$ be a complex projective plane referred to homogeneous coordinates x,y,z. That is to say, a point M is any triple x,y,z of numbers not all zero and proportional triples represent the same point. Let continuity in $\mathfrak{P}$ be defined about each point M by the continuity of the ratios of the coordinates to one of them which is $\neq 0$ at M. Show that with this topology $\mathfrak{P}$ is an arc-wise connected polyhedron and calculate its Betti numbers, likewise its fundamental and homotopy groups.

5. In the notations of the preceding problem calculate the fundamental groups of the complement in $\mathfrak{P}$ of the locus (algebraic curve) $y^2z = x^3$.

6. Let T_n denote the product of n circumferences (n-dimensional torus). Show that T_n is a polyhedron and calculate its Betti numbers. Show that the fundamental group of T_n is a free commutative group on n generators and calculate its homotopy groups.

7. Let T,T' be two solid tori and Φ,Φ' their surfaces. Let α,β be the fundamental one-cycles on a meridian and a parallel of Φ, and let α',β' be the same for Φ'. Let f be a homeomorphism between Φ and Φ' and let corresponding points be identified, thus giving rise to a surface Ψ and a solid $\Omega = T \cup T'$. We suppose f such that Ω is a polyhedron with Ψ as a subpolyhedron. On the surface Ψ we will have

$$\alpha' \sim p\alpha + q\beta, \qquad \beta' \backsim r\alpha + s\beta,$$
$$ps - qr = \pm 1.$$

Conversely if p,q,r,s are any integers satisfying the relation just written, Ω may be constructed in the manner described. Calculate the Betti numbers and the fundamental group of Ω. The polyhedron Ω is known as a *lense-space* (Seifert-Threlfall).

8. Two lense-spaces may have the same fundamental group without being homeomorphic (J. W. Alexander).

9. Let S^2 be a topological sphere in Euclidean three-space, and Ω the region bounded by S^2. When S^2 is analytical, Ω is a three-cell. Construct an S^2 such that Ω has an infinite fundamental group and hence is not a three-cell (J. W. Alexander).

Remark. The surface Φ in $\mathfrak{E}^3$ is said to be *analytical* whenever every point has a neighborhood in which one of the coordinates is an analytical function of the other two.

10. Let the analytical surface Φ of genus $p > 0$ in $\mathfrak{E}^3$ bound a region Ω. The set $\Omega \cup \Phi$ is a polyhedron Π. Calculate the Betti numbers of Π, likewise the groups $\pi_n(\Pi)$.

11. A metric space $\mathfrak{R}$ is said to be an LC^n space whenever given any point x of $\mathfrak{R}$ and any $\epsilon > 0$ there is an $\eta > 0$ such that every mapping φ of a sphere S^k, $k \leq n$, into $\mathfrak{S}(x,\eta)$ is homotopic to a constant in $\mathfrak{S}(x,\epsilon)$.

Let $\mathfrak{R}$ be a connected compactum which is an LC^n, $n > 0$. Show that $\mathfrak{R}$ is arc-wise connected. Show also that its groups $\pi_h(\mathfrak{R})$, $h \leq n$, likewise a universal covering space $\mathfrak{S}$, may be defined and prove that $\pi_h(\mathfrak{R}) = \pi_h(\mathfrak{S})$, $1 < h \leq n$.

12. Let $\mathfrak{R}$ be arc-wise connected and let $P \in \mathfrak{R}$. Consider a disk Δ^n with boundary sphere S^{n-1}. Then $\pi_n(\mathfrak{R})$ may be defined by means of the mapping $\alpha : \Delta^n \to \mathfrak{R}$ such that $\alpha S^{n-1} = P$, as in the text by means of punctured spheres. Prove that the two definitions are actually equivalent.

13. If Π and Ω are polyhedra, then the components of function space Π^Ω are arc-wise connected.

14. Let Φ be the subset of Π^{S_n} consisting of the mappings $S^n \to \Pi$ sending a given point Q of S^n into the base point P of Π. The components of Φ are arc-wise connected. Let Λ be the component containing the constant mapping $S^n \to P$. Assuming $n \geq 1$ we have then $\pi_{n+1}(\Pi) = \pi_1(\Lambda)$. This provides a direct definition of the groups of π_n in terms of suitable groups π_1. Prove directly by means of this definition that the groups π_p, $p > 1$, are all commutative.

VI. Introduction to Manifolds

Manifolds have been mentioned more than once in the preceding pages, and those of dimension two have been rather fully treated in Chapter II. In the present chapter we shall consider a relatively elementary type introduced in substance by Poincaré and prove his duality theorem for such manifolds.

§1. Differentiable and Other Manifolds.

1. An *n-dimensional manifold* M^n, as commonly understood, is first a topological space such that each point x has for neighborhood an n-cell $E^n(x)$. In addition there are usually imposed some properties which guarantee the existence of "continuously turning" tangent lines (if M is a curve) or tangent spaces or else normal spaces. These properties cannot be defined unless the cells are implemented with parametrizing variables possessing suitable differentiability properties. These terms will be clarified in a moment. Examples of manifolds such as we have in view are the "smooth" nonsingular curves and surfaces of algebraic geometry, Euclidean and real or complex projective spaces, etc.

The manifolds which have merely cells for neighborhoods without any other specific properties, are known as *topological*. Their status is still very obscure. Thus it is not even known whether a compact topological manifold is a polyhedron, one of the well known unsolved problems proposed by Poincaré.

It so happens that as soon as one imposes reasonable differentiability the manifolds become polyhedra and thus topologically at least when they are compact, they fall within the scope of our program. Our first objective is to describe the mode of linking of the differentiability properties with topology.

2. We begin by stating a few necessary mixed concepts involving both differentiation and topology.

A finite set $\{f_i(u_1, \cdots, u_n)\}$ of real functions defined in a region Ω of the Euclidean space U^n of $u_1, \cdots, u_n$ is said to be *regular* when the f_i and the $\partial f_i/\partial u_j$ exist and are continuous in Ω and the Jacobian matrix $||\partial f_i/\partial u_j||$ is of maximum rank at every point of Ω.

A *parametric n-cell* is a triple (E^n,t,U^n) where E^n is an n-cell, U^n is as above and t is a topological mapping of $\bar{E}^n$ into a closed spherical region of U^n. The u_i are then called the *parameters* of the cell.

A parametric p-cell (E^p, s, V^p) is said to be *regularly imbedded* in (E^n,t,U^n) whenever $E^p \subset E^n$ and on E^p the v-coordinates of the points form a regular set of functions of their u-coordinates. We shall also say briefly: the first is a *p-subcell* of the second.

3. We may now define a *differentiable n-manifold* M^n as follows: It is a topological space with a countable open covering by parametric n-cells $\{(E_i^n,t_i,U_i^n)\}$ referred to as a *parametric covering,* such that: (a) the covering is *locally finite,* i.e. any point x of M^n is in at most a finite number of cells E_i^n; (b) if x is in both E_i^n and E_j^n then the i-th and j-th parametric cells have in common an n-subcell containing x.

It is clear that M^n is a topological n-manifold.

Any subcell of a parametric n-cell of the parametric covering of M^n is called a *subcell* of M^n.

The differentiable manifold M^p is said to be *regularly imbedded* in the differentiable manifold M^n whenever $M^p \subset M^n$ and any point of M^p is contained in a p-subcell of M^p which is also a subcell of M^n.

If M^n and M'^n are the same topological space and one of them is regularly imbedded in the other, then the reverse is also true and the two manifolds are considered as *identical.*

Examples. A coordinate system of an Euclidean n-space $\mathfrak{E}^n$ defines it as a differentiable M^n. This manifold is independent of the linear transformation of coordinates. A similar remark with unimportant modifications may be made for projective spaces. Nonsingular algebraic surfaces and the manifolds of the theory of relativity are all differentiable.

Whenever M^n is compact the covering $\{(E_i^n,t_i,U_i^n)\}$ may be chosen finite and no mention of local finiteness is then necessary.

4. (4.1) **Theorem.** *Every compact differentiable M^n may be regularly imbedded in some Euclidean space.*

The theorem was first proved by Whitney [a] and even without the restriction to compact manifolds. However as we do not expect to consider other manifolds the restriction will not matter and is well worth the simplification in the proof.

Let first (E_o^n,t,U^n) be any subcell of M^n and let $x \in E_o^n$, $y = tx$. Then there exists a sphere $\mathfrak{S}(y, \rho)$ such that $\bar{\mathfrak{S}}(y, 4\rho)$ is contained in tE_o^n. Thus if $E_o'^n = t^{-1}\, \mathfrak{S}(y, \rho)$, x is contained in a parametric cell $(E_o'^n,t,U^n)$ such that $tE_o'^n = \mathfrak{S}(y, \rho)$ and that t is defined on a closed cell $\bar{E}_o''^n \supset E_o'^n$ such that $tE_o''^n = \mathfrak{S}(y, 4\rho)$. Since M^n is compact it has a finite parametric covering $\{(E_i^n,t_i,U_i^n)\}$ with the following property.

For every i there are cells $E_i'^n$, $E_i''^n$ such that $E_i^n \subset E_i'^n \subset E_i''^n$ and that t_i is topological on $\bar{E}_i''^n$ to U_i^n, and also that $t_iE_i^n = \mathfrak{S}(y_i, \rho_i)$, $t_iE_i'^n = \mathfrak{S}(y_i, 2\rho_i)$, $t_iE_i''^n = \mathfrak{S}(y_i, 4\rho_i)$.

Define now a real nonincreasing function $\varphi(z)$ continuous together with its first derivative on $0 \leq z \leq 1$ and such that

$$\varphi(z) = 1,\ 0 \leq z \leq \tfrac{1}{2};\ \varphi(1) = 0.$$

Set also $r_i = d(t_ix, y_i)$ for $x \in \bar{E}_i''^n$. Finally if u_{ih}, $h = 1, 2, \cdots, n$ are the coordinates in U_i^n, set

$$v_{ih} = \varphi\left(\frac{r_i}{2\rho_i}\right)(u_{ih} + 4\rho_i),\ x \in E_i'^n,$$

$$v_{ih} = 0,\ x \in M^n - E_i'^n,$$

$$v_{i,n+h} = \varphi\left(\frac{r_i}{4\rho_i}\right)(u_{ih} + 4\rho_i),\ x \in E_i''^n,$$

$$v_{i,n+h} = 0,\ x \in M^n - E_i''^n.$$

The following properties may now be noted:

(a) The set $\{v_{ih}, v_{i,n+h}\}$ is regular in each $E_i'^n$ and continuous on M^n.

(b) On the respective cells we have:

$$E_i^n : v_{ih} = u_{ih} + 4\rho_i;$$

$$E_i'^n : v_{i,n+h} = u_{ih} + 4\rho_i,\ v_{ih} > 0;$$

$$\text{outside of } E_i'^n : v_{ih} = 0;$$

$$h = 1, 2, \cdots, n.$$

Let the number of cells E_i^n be q and set $p = 2nq$. Consider now the v_{ih}, $v_{i,n+h}$ as coordinates for an Euclidean space $\mathfrak{E}^p$ and assign to any point x of M^n the point X of $\mathfrak{E}^n$ whose coordinates are the $v_{ih}(x)$, $v_{i,n+h}(x)$. Owing to (a), $f : x \to X$ defines a mapping $f : M^n \to \mathfrak{E}^p$. Let $\mathfrak{M}^n = fM^n$, and x_1, x_2 be two distinct points of M^n. Since $\{E_i^n\}$ is a covering of M^n one of its sets say E_i^n contains x_1. Regarding x_2 there are two possibilities:

I. *x_2 is in $E_i'^n$.* Then x_1, x_2 have different coordinates u_{ih} and hence different $v_{i,n+h}$. Therefore $fx_1 \neq fx_2$.

II. *x_2 is not in $E_i'^n$.* Then the $v_{ih}(x_1)$ are positive and the $v_{ih}(x_2)$ are zero. Hence again $fx_1 \neq fx_2$.

Thus f is a univalent mapping $M^n \to \mathfrak{M}^n$. Since M^n is compact f is topological. Since $\{v_{ih}, v_{i,n+h}\}$ is regular in each E_i^n, the imbedding is differentiable. This proves the theorem.

5. It has been shown by S. S. Cairns [a] that

(5.1) **Theorem.** *Every compact differentiable manifold M^n in an Euclidean space $\mathfrak{E}^q$ may be covered with a simplicial complex. Thus a compact differentiable M^n is a polyhedron.*

The proof of this theorem would take us too far afield and so it is omitted. Let us merely state that a proof of the same theorem for non-compact manifolds has been given by J. H. C. Whitehead [a]. The latter has even established an important complementary property. Let $K = \{\sigma\}$ be a covering complex and let $St\ \sigma = \{\sigma\sigma'\}$. Then $Lk\ \sigma = \{\sigma'\}$ is a complex. For if σ'' is a face of σ' then $\sigma\sigma'' \in St\ \sigma$ and so $\sigma'' \in Lk\ \sigma$. We refer to $Lk\ \sigma$ as the *linked complex* of σ. And now the following result is implicit in Whitehead's proof:

(5.2) *The covering complex K of M^n may be chosen such that every $Lk\ \sigma^p$ is an $(n - p - 1)$-sphere. As a consequence every star-set $|St\ \sigma|$ is an n-cell. Furthermore a complex such as K may be chosen whose mesh is arbitrarily small.*

In the present chapter only complexes possessing property (5.2) are to be considered. As they are actually the manifolds investigated by Poincaré we may formulate the explicit

Definition. A *Poincaré n-manifold* is a connected complex whose linked complexes $Lk\ \sigma^p$ are $(n - p - 1)$-spheres.

Instead of "Poincaré manifold" we shall usually merely say "manifold."

It is easy to verify that a subdivided Jordan curve and a closed surface in the sense of (II) are Poincaré manifolds.

6. (6.1) *A Poincaré manifold is an n-circuit.*

Let $K = \{\sigma\}$ and let $\Lambda = \{\sigma_i^{n-1}, \sigma_j^n\}$. According to (III, 21) the following three properties must be verified:

(a) Every σ^{n-1} of K is a face of exactly two σ^n's.

(b) Λ is connected.

(c) No proper subcomplex of K possesses properties (a) and (b).

Since $Lk\ \sigma^{n-1}$ is a zero-sphere, it consists of two points and this implies (a).

Since $Lk\ \sigma^p$ is an $(n - p - 1)$-sphere it contains a σ^{n-p-1}. Hence $St\ \sigma^p$ contains the n-simplex $\sigma^p\sigma^{n-p-1}$ or σ^p is a face of a σ^n. Hence every simplex of K is a face of a simplex of Λ.

Suppose now Λ not connected and let Λ_1 be one of its components and $\Lambda_2 = \Lambda - \Lambda_1$. Denote by L_i the set of faces of the simplexes of Λ_i. Then $K = L_1 \cup L_2$. Since K is connected L_1 and L_2 have common simplexes. Let σ^p be one of the highest dimension common to both and let P_i be the set of elements of L_i in $Lk\ \sigma^p$. Under our assumption $p < n - 1$, hence dim $Lk\ \sigma^p > 0$ and so $Lk\ \sigma^p$ is connected. On the other hand $Lk\ \sigma^p = P_1 \cup P_2$ and P_1, P_2 are disjoint since if σ were a common element L_1 and L_2 would have the common simplex $\sigma\sigma^p$ whose dimension exceeds p. This contradiction proves that Λ is connected, which is (b).

Suppose now that L, a subcomplex of K, possesses properties (a) and (b). Then L must contain n-simplexes and in view of (a), (b) it must contain all of Λ. Every $\sigma \in K$ must be in L and so $L = K$. Thus (c) is also verified. This proves (6.1).

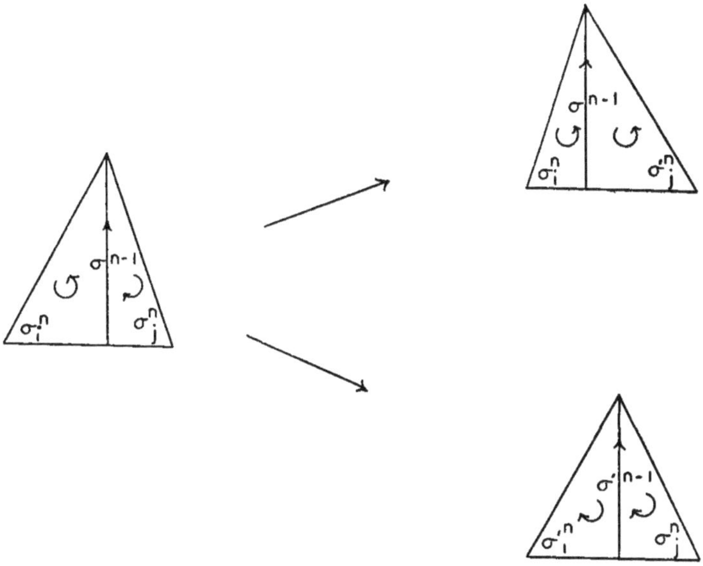

Figure 70

The circuit properties of K are "transferred" to M^n. We shall say that M^n is *orientable* or *oriented* whenever K has one or the other property.

Convention. Henceforth M^n is assumed compact, connected, orientable and oriented (wherever necessary). The nonorientable manifolds may always be dealt with by restricting the discourse to chains, · · · , mod 2. This will make it unnecessary to deal with them separately.

There is a noteworthy construction, due to de Rham, associated with the question of orientability. Take the simplexes of K and for each σ introduce a new one σ' oriented like σ if dim $\sigma < n$, oriented oppositely if dim $\sigma = n$. The new collection $\{\sigma,\sigma'\}$ is now reassembled to form a simplicial complex $\mathfrak{K}$ in the following manner. Let σ^{n-1} be a common face of σ_i^n, σ_j^n. If σ^{n-1} is oppositely related to σ_i^n, σ_j^n we match them across σ^{n-1} as before and match $\sigma_i'^n$, $\sigma_j'^n$ across σ'^{n-1}. If σ^{n-1} is oriented alike as to σ_i^n, σ_j^n we match σ_i^n and $\sigma_j'^n$ across say σ^{n-1} and $\sigma_i'^n$, σ_j^n across σ'^{n-1} (see Fig. 70). The simplexes of dimension $< n - 1$ are matched in such manner as to obtain a simplicial complex $\mathfrak{K}$. It is evident from the construction that every component of $\mathfrak{K}$ is an orientable combinatorial n-manifold. If K is orientable $\mathfrak{K}$ has two components, each an isomorph of K. However if K is nonorientable $\mathfrak{K}$ is itself an orientable M^n.

Assuming then K nonorientable, $\mathfrak{K}$ determines an orientable manifold $\mathfrak{M}^n$ with the following property: there exists a so-called $(2 - 1)$ mapping $\varphi : \mathfrak{M}^n \to M^n$; each point of M^n is the image of two points of $\mathfrak{M}^n$. We refer to $\mathfrak{M}^n$ as a *doubly covering manifold* for M^n. It is fairly clear that the study of M^n may be replaced by that of $\mathfrak{M}^n$.

§2. THE POINCARÉ DUALITY THEOREM.

7. This theorem, the first duality theorem discovered in topology, was given by Poincaré in 1895. Its statement is as follows:

(7.1) *The Poincaré duality theorem. The Betti numbers of an orientable manifold M^n satisfy the relations*

(7.2) $$R_\pi^p = R_\pi^{n-p} \; (\pi = 0 \textit{ for rational cycles}).$$

It should be added that Poincaré only gave the proof for rational cycles. The proof for mod 2 was given by Veblen [V] and for mod π by Alexander [c].

As a complement we may add:

(7.3) *When M^n is nonorientable one may only assert that*

(7.4) $$R_2^p = R_2^{n-p}.$$

In other words only the mod 2 *part of the theorem remains.*

The treatment of the nonorientable M^n is essentially the same as for the orientable case, except that only chains mod 2 are admitted everywhere. The modifications being obvious, it will not be necessary to consider nonorientable manifolds.

Method of proof. We shall introduce in relation to the covering complex K of (5.2) a certain general complex $\bar{K}$, the *reciprocal* of K, and show that it has the same derived K' as K. Moreover it will also be shown that $\bar{K}$ is related to K' like K. The comparison of the homology groups of the three complexes will then yield the duality theorem. The crux of the whole argument is found in the relation of $\bar{K}$ to K'. Once this is disposed of the proof will be immediate.

8. To understand the relation of the complex K to its reciprocal it will be helpful to examine certain well known configurations.

Consider a regular tetrahedron $\sigma^3 = ABCD$ and let $K = \mathfrak{B}\sigma^3$ be the triangulated boundary two-sphere. Through each edge pass the plane perpendicular to the opposite edge, (Fig. 71). The six planes thus obtained subdivide the sphere into the elements of the barycentric derived K'. In Fig. 72 the elements of the star of A in K' are shown "flattened out" on a plane. The following facts are readily verified:

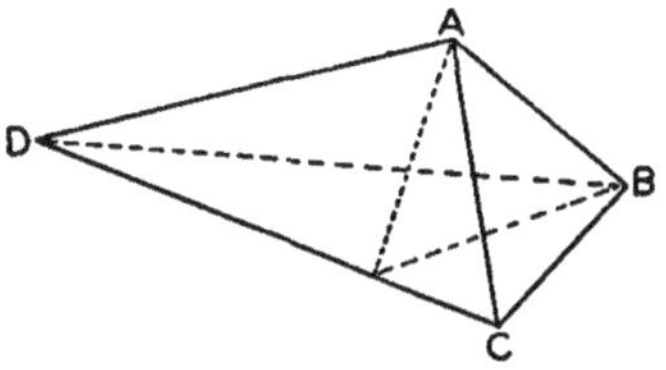

Figure 71

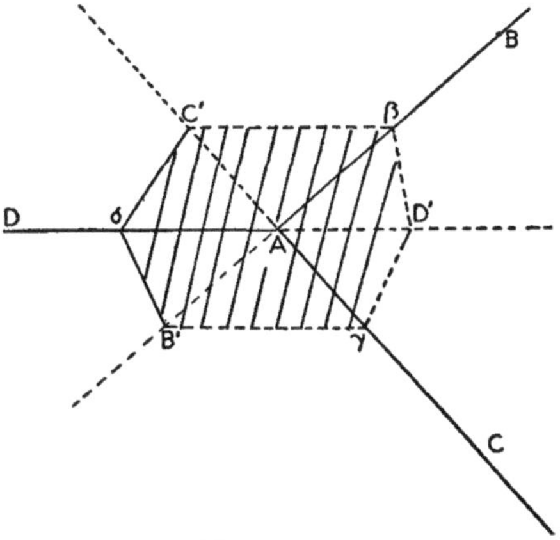

Figure 72

(a) to each vertex of K, such as A, there corresponds a two-cell such as the shaded hexagonal region in the figure; there are four such two-cells;

(b) to each edge of K, such as AD, there corresponds a "transverse" open one-cell such as $C'\delta B'$; there are six one-cells of this kind;

(c) to each triangle of K, for instance ABC, there corresponds a vertex, here D'; there are four vertices of this nature.

Let $\{\sigma_i^p\}$ denote the elements of K. The cell associated in the above scheme with σ_i^p is of dimension $2 - p$ and we denote it by ξ_i^{2-p}. It is clear that the ξ's make up a decomposition of the boundary sphere into cells. It is also readily verified that the correspondence $\sigma \leftrightarrow \xi$ reverses the "is a face of" relations. In point of fact here $\bar{K} = \{\xi\}$ has exactly the same structure as K itself, i.e. it is a simplicial complex. In short by regrouping the elements of K' in a certain way we have obtained a new complex $\bar{K}$ associated with K in an "incidence-reversing" manner.

Something quite similar may be verified regarding the boundary of any convex polyhedral region in ordinary space, or the triangulation of any closed surface (in the sense of II, 17). Generally, however, the new cells thus arising will fail to be simplicial. Thus the subdivision of a sphere by three mutually perpendicular planes through the center gives rise to some cells ξ_i^2 associated with the vertices whose appearance is that of an octagonal region. Furthermore for higher dimensions a precisely similar situation prevails in which if the dimension of the complex is n the associated elements are σ_i^p and ξ_i^{n-p}. Thus if a cubical

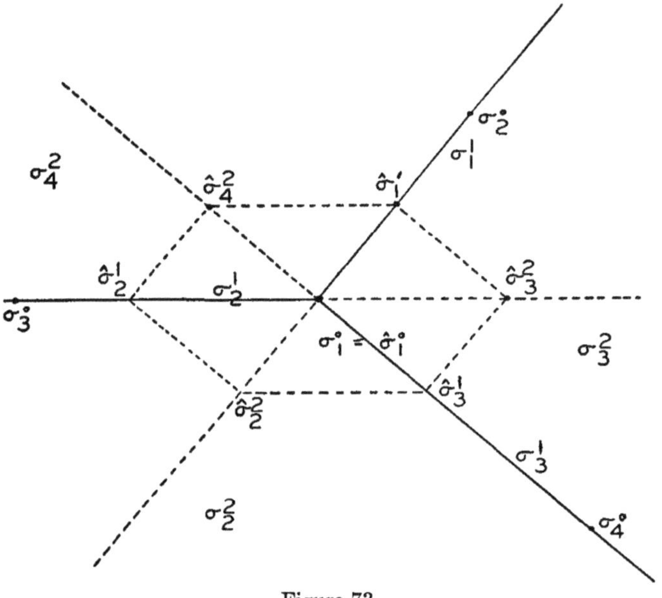

Figure 73

lattice is subdivided barycentrically the elements of the subdivision clustering around a vertex A of the lattice make up a three-cell. The reversal of the incidences is also absolutely general.

For a better understanding of the situation, it is advisable to introduce our standard "σ, $\hat{\sigma}$" designation for the simplexes of K and their centroids. Thus Fig. 73 is merely Fig. 72 with the elements relabeled in the way just indicated. Remembering the designations $\zeta = \hat{\sigma}_i \cdots \hat{\sigma}_j$, $\sigma_i < \cdots < \sigma_j$ for the simplexes of K' and the terms "first vertex" for $\hat{\sigma}_i$ and "last vertex" for $\hat{\sigma}_j$, we verify the following rule: the cell ξ_i^{2-p} consists of all the simplexes ζ of K' which begin with $\hat{\sigma}_i^p$. Thus the hexagonal two-cell ξ_1^2 consists of all the simplexes which begin with

$\hat{\sigma}_1^0 : \hat{\sigma}_1^0, \hat{\sigma}_1^0\hat{\sigma}_2^2, \hat{\sigma}_1^0\hat{\sigma}_2^1\hat{\sigma}_4^2, \cdots$, This is the property which will serve as a basis for our definition of the reciprocal complex.

9. The preceding considerations are now to be extended to any manifold M^n of our restricted class: compact, connected and orientable. The manifold is supposed to be covered with a certain simplicial complex $K = \{\sigma\}$ which satisfies (5.2).

Corresponding to any σ_i^p let ξ_i^{n-p} represent the point-set union of all the simplexes of the derived K' which begin with $\hat{\sigma}_i^p$, and let us compare ξ_i^{n-p} with $|Lk\ \sigma_i^p|$.

The derived $(Lk\ \sigma_i^p)'$ consists of all the simplexes of K' of the form

$$\zeta = \hat{\sigma}_j \cdots \hat{\sigma}_k, \qquad \sigma_i^p\sigma_j < \cdots < \sigma_i^p\sigma_k.$$

On the other hand ξ_i^{n-p} consists of all the simplexes of K' represented by

$$\zeta_1 = \hat{\sigma}_i^p\ \widehat{\sigma_i^p\sigma_j} \cdots \widehat{\sigma_i^p\sigma_k}, \qquad \sigma_i^p\sigma_j < \cdots < \sigma_i^p\sigma_k,$$

where $\widehat{\sigma\sigma'}$ denotes the vertex of K' in the simplex $\sigma\sigma'$ of K. Hence if

$$\zeta_2 = \widehat{\sigma_i^p\sigma_j} \cdots \widehat{\sigma_i^p\sigma_k},$$

then $\Lambda = \{\zeta_2\}$ is a simplicial complex isomorphic with $(Lk\ \sigma_i^p)'$, the isomorphism being defined by $\hat{\sigma}_j \leftrightarrow \widehat{\sigma_i^p\sigma_j}$, $\sigma_i^p < \sigma_j$. Hence (5.2) Λ is an $(n - p - 1)$-sphere. Since ξ_i^{n-p} consists of the simplexes of the join $\hat{\sigma}_i^p\Lambda$, ξ_i^{n-p} is an $(n - p)$-cell.

The boundary sphere $|\Lambda|$ of the cell ξ_i^{n-p} is the point-set union of all the simplexes of K' which begin with a vertex $\hat{\sigma}$ such that σ_i^p is a proper face of σ. Hence $|\Lambda|$ is the point-set union of all the ξ_j^{n-q} such that σ_j^q is a proper face of σ_i^p. Let us call ξ_i^{n-p} and any ξ_j^{n-q} a *face* of ξ_i^{n-p}. The collection of all the faces of ξ_i^{n-p} is called the *closure* of ξ_i^{n-p}, written $Cl\ \xi_i^{n-p}$. The set of all the *proper faces* of ξ_i^{n-p} (faces other than ξ_i^{n-p}) is called the *boundary* of ξ_i^{n-p} written $\mathfrak{B}\xi_i^{n-p}$. If A is any collection of ξ's then $|A|$ will designate their point-set union. In particular $|\mathfrak{B}\xi^q|$ is a $(q - 1)$-sphere and $|Cl\ \xi^q|$ is a q-cell.

All this is very natural and follows most closely what has been done for simplicial complexes. To complete the picture, we assign to ξ^q the natural "arithmetical" *dimension* q and define *incidence numbers*

$$[\xi_i^{n-p} : \xi_j^{n-p-1}] = [\sigma_j^{p+1} : \sigma_i^p]. \tag{9.1}$$

The relation between the collection $\bar{K} = \{\xi\}$ and K is then as follows: $\sigma_i^p \leftrightarrow \xi_i^{n-p}$ is one-one; it reverses the incidences $<$; it does not modify the incidence numbers and associated elements have dimensions whose sum is n.

It follows from the properties just proved and (III, 3.1) that as in simplicial complexes:

$$\sum_i [\xi^q : \xi_i^{q-1}][\xi_i^{q-1} : \xi^{q-2}] = 0. \tag{9.2}$$

Thus the collection $\bar{K}$ is implemented with the same features as any simplicial complex. We refer to $\bar{K}$ as the *reciprocal* complex of K and observe at once that all the algebraic properties developed for K and its chains, cycles, $\cdots$, may be immediately extended to $\bar{K}$. The only exceptions are the properties of connectedness which we shall not need.

It is convenient to point out that if the boundary relations in K are

$$F\sigma_i^{p+1} = \sum_j \eta_{ij}(p)\sigma_j^p, \tag{9.3}$$

then those in $\bar{K}$ are

$$F\xi_j^{n-p} = \sum_i \eta_{ij}(p)\xi_i^{n-p-1}. \tag{9.4}$$

(We also note explicitly that in $\bar{K} : FF = 0$.)

If we wrote the boundary relations of $\bar{K}$ in the same form as (9.3) we would have

$$F\xi_i^{p+1} = \sum \bar{\eta}_{ij}(p)\xi_j^p \tag{9.5}$$

and the relation between the incidence matrices $\eta(p)$, $\bar{\eta}(p)$ is

$$^{-}(n - p - 1) = \eta(p). \tag{9.6}$$

An important observation may be made regarding $\bar{K}$. Since K is an orientable n-circuit it may also be oriented. Let this be assumed once for all. Under the circumstances every σ_i^{n-1} will be adjacent to precisely two n-simplexes σ_j^n and σ_k^n and oppositely related to them. Hence every ξ_i^1 has for boundary two vertices ξ_j^0, ξ_k^0 and is oppositely related to them. Therefore the one-section $\bar{K}^1$ of $\bar{K}$ (its set of cells of dimension zero and one) is a simplicial complex.

A subcollection L of $\bar{K}$ is called a *closed subcomplex* of $\bar{K}$ whenever it contains the closures of its own elements.

If $N = \{\sigma_i'^p\}$ is any subcollection of K then $\{\xi_i'^{n-p}\}$ written $\bar{N}$ is called the *reciprocal* of N.

If $P = \{\xi\}$ is any subcollection of $\bar{K}$ then the point-set union $\bigcup \xi$ is written $|P|$.

10. The construction of the derived $\bar{K}'$ may be carried out as for simplicial complexes. It is natural, however, to select as new vertex in ξ_i^{n-p} the centroid $\hat{\sigma}_i^p$ of σ_i^p. Under our rules the elements of $\bar{K}'$ are to be represented by expressions

(10.1) $\quad \zeta_o = \hat{\sigma}^p_i \cdots \hat{\sigma}^q_j, \qquad \xi^{n-p}_i < \cdots < \xi^{n-q}_j.$

Since the incidences $<$ are reversed from K to $\bar{K}$, ζ_o is the same, except for orientation, as the simplex of K' given by

(10.2) $$\zeta = \hat{\sigma}^q_j \cdots \hat{\sigma}^p_i.$$

There is an obvious one-one correspondence $\zeta_o \leftrightarrow \zeta$ and therefore $\bar{K}' \cong K'$. Thus the complex K and its reciprocal $\bar{K}$ have the same derived. Henceforth we shall continue to adhere to the designation (10.2) for the simplexes of K'.

In the treatment of derivation in $\bar{K}$ we shall require the following property:

(10.3) $$|Cl\ \xi^{n-p}_i| \subset |St\ \sigma^p_i|.$$

The set at the left is the union of all the simplexes ζ of K' which begin with a $\hat{\sigma}_j$ where $\sigma_j \in St\ \sigma^p_i$. Since any such ζ is in $|St\ \sigma_j| \subset |St\ \sigma^p_i|$, (10.3) follows.

11. Continuing with our basic program we shall now develop for K', as the derived of $\bar{K}$, operations $\bar{D}$, $\bar{\delta}$, $\bar{\tau}$ wholly analogous to the operations D, δ, τ of (IV, 3, 5) for K, K'. The (future) isomorphisms induced by $\bar{\tau}$, $\bar{\delta}$ will be written $[\bar{\tau}]$, $[\bar{\delta}]$. Formally speaking, we shall write $\hat{\xi}^{n-p}_i$, or merely $\hat{\xi}$, for $\hat{\sigma}^p_i$.

The operations $\bar{D}$, $\bar{\delta}$ are readily defined by analogy with the earlier situation:

(11.1) $\quad \bar{D}\xi^o = \hat{\xi}^o = \xi^o; \qquad \bar{D}\xi^p = \hat{\xi}^p \cup \hat{\xi}^p\bar{D}\mathfrak{B}\xi^p, \qquad p > 0;$

(11.2) $\quad \bar{\delta}\xi^o = \hat{\xi}^o = \xi^o; \qquad \bar{\delta}\xi^p = \hat{\xi}^p\bar{\delta}F\xi^p, \qquad p > 0.$

The proof that $\bar{\delta}$ is a chain mapping $\bar{K} \to K'$ proceeds as in (IV, 5) and need not be repeated.

Let $\bar{K}^p$ denote the p-section of $\bar{K}$. Since generally only K^1 is simplicial and $\bar{K}$ itself is not, the definition of $\bar{\tau}$ will not be as immediate as in (IV, 5).

We first define an important set transformation $t : K' \to \bar{K}$ (a future carrier) as follows: If $\zeta = \hat{\sigma}^p_i \cdots \hat{\sigma}_j$ $(\sigma^p_i < \cdots < \sigma_j)$ then $t\zeta = Cl\ \xi^{n-p}_i$. If $\zeta' = \hat{\sigma}^q_h \cdots \hat{\sigma}_k < \zeta$ $(\sigma^q_h < \cdots < \sigma_k)$ then $\sigma^p_i < \sigma^q_h$ and hence $\xi^{n-q}_h < \xi^{n-p}_i$. Therefore $t\zeta' = Cl\ \xi^{n-q}_j \subset Cl\ \xi^{n-p}_i = t\zeta$.

(11.3) *There may be defined a chain-mapping $\bar{\tau} : K' \to \bar{K}$ with the following properties:*

(a) *$\bar{\tau}$ has the carrier t and this carrier is zero-cyclic;*

(b) *$\bar{\tau}\bar{\delta} = 1$;*

(c) *$\bar{\delta}\bar{\tau} \backsim 1$ in t.*

It will then be a consequence of the last two properties that

(d) $\bar{\delta}$, $\bar{\tau}$ *induces isomorphisms of the homology groups of* $\bar{K}$ *with the corresponding groups of* K', *and* $[\bar{\tau}] = [\bar{\delta}^{-1}]$.

Let $(11.3)_p, \cdots, (c)_p$ designate the fact that $\bar{\tau}$ has already been defined on the p-section $\bar{K}^p$ and that (a), (b), (c) hold as regards operations in $\bar{K}^p$. Since the operations under consideration send $\bar{K}^p$ into $(\bar{K}^p)'$ and $(\bar{K}^p)'$ into $\bar{K}^p$, it will then follow that (d) holds as regards $\bar{K}^p$ instead of $\bar{K}$. More generally if $\bar{L}$ is a subcomplex of $\bar{K}$ then $L' = \bar{L}'$ (the set of the derived of the cells of $\bar{L}$) is a subcomplex of K' and $tL' = \bar{L}$. From this follows that if dim $\bar{L} \leqq p$ then $(11.3)_p$ implies that.

(11.4) $\bar{\delta}$ *(operating in* $\bar{L}$*) induces an isomorphism of the homology groups of* $\bar{L}$ *with the corresponding groups of* L' *and* $[\bar{\tau}] = [\bar{\delta}^{-1}]$.

Let $\bar{\mathfrak{D}}$ denote the homotopy operator for (c). Since $tL' = \bar{L}$, if (c) holds then $\bar{\mathfrak{D}}L' \subset \bar{L}$. Hence if $(11.4)_p$ denotes (11.4) for dim $\bar{L} \leqq p$, then $(11.3)_p$ will imply $(11.4)_p$, and in particular the groups of $\bar{L}$, L' will be isomorphic when dim $\bar{L} \leqq p$.

Since $\bar{K}^1$ is simplicial we prove $(11.3)_1$ as in the simplicial case. Hence we may assume $(11.3)_{p-1}$, $p > 1$, and prove $(11.3)_p$. Once we have $(a)_p$ the proofs of $(b)_p$, $(c)_p$ proceed as in (IV, 5) and (V, 6) and need not be repeated. Thus only $(a)_p$ requires special attention.

Since $(11.3)_{p-1}$ holds $\mathfrak{B}\xi_j^p$ and $(\mathfrak{B}\xi_j^p)'$ have the same homology groups. Since $(\mathfrak{B}\xi_j^p)'$ is merely a simplicial complex covering a $(p-1)$-sphere, its homology groups are those of the latter and so are those of $\mathfrak{B}\xi_j^p$. Since $p \geqq 2$, $\mathfrak{B}\xi_j^p$ is cyclic in the dimensions 0, $p-1$ and acyclic in the rest.

Consider now $Cl\ \xi_j^p$. Since it has the same homology groups as $\mathfrak{B}\xi_j^p$ for the dimensions $< p - 1$, it is cyclic in the dimensions zero and acyclic in every dimension q such that $0 < q < p - 1$. Since $\mathfrak{B}\xi_j^p$ is cyclic in the dimension $p - 1$, which is its own dimension, its $p - 1$ cycles are all of the form $m\gamma^{p-1}$, where γ^{p-1} is a fixed cycle of the sphere. On the other hand

$$\gamma'^{p-1} = F\xi_j^p = \Sigma[\sigma_k^{n-p+1}:\sigma_j^{n-p}]\xi_k^{p-1},$$

where the sum is extended to the faces of ξ_j^p, is a cycle of $\mathfrak{B}\xi_j^p$. We have then $\gamma'^{p-1} = \mu\gamma^{p-1}$ where μ is an integer. Since the incidence numbers [:] are ± 1, necessarily $\mu = \pm 1$. Hence $\gamma^{p-1} \sim 0$ in $Cl\ \xi_j^p$ and $Cl\ \xi_j^p$ is acyclic in the dimension $p - 1$. Its p-chains are all of the form $m\xi_j^p$, and have a boundary $m\gamma'^{p-1} \neq 0$. Hence $Cl\ \xi_j^p$ is acyclic in the dimension p also, and consequently in all dimensions $q > 0$. Thus $Cl\ \xi_j^p$ is zero-cyclic, and so t behaves in accordance with $(a)_p$.

Since $\bar{K}^1$ is simplicial, $\bar{\tau}$ is simplicial on this complex, and $t|\bar{K}^p$ is zero-cyclic. As a consequence $\bar\tau$ may be extended to $\bar K$ with the carrier t (V, 5.2). This completes the proof of $(a)_p$ and hence also the proof of (11.3).

12. It is a consequence of (11.3) that

$$\mathfrak{H}^p(\bar K,G) \cong \mathfrak{H}^p(K',G) \cong \mathfrak{H}^p(K,G).$$

Hence $\bar K$ has the same homology groups as K itself.

On the other hand let us calculate the Betti number $\bar R^q_\pi$ of $\bar K$ directly as in (III, 16).

If $\bar\alpha^q$, $\bar\rho^q_\pi$ denote the number of $\bar\xi^q_i$ and the rank of $\bar\eta(q)$ mod π we have by (III, 16.8), and recollecting that $\bar\alpha^{n-q} = \alpha^q$, the number of σ^q_i, and $\bar\rho^{n-q}_\pi = \rho^q_\pi$:

$$R^{n-p}_\pi = \bar R^{n-p}_\pi = \bar\alpha^{n-p} - \bar\rho^{n-p-1}_\pi - \bar\rho^{n-p}_\pi = \alpha^p - \rho^p_\pi - \rho^{p-1}_\pi = R^p_\pi.$$

Thus (7.2) holds for K and hence for M^n. This completes the proof of Poincaré's duality theorem.

13. Extensions. I. Let us drop the condition that M^n be connected. Let $\{K_i\}$ be the components of K. As we know, if $\{M^n_i\}$ are the components of M^n property labeled, there is a one-one correspondence $M^n_i \leftrightarrow K_i$ such that K_i covers M^n_i. Each M^n_i is then a connected manifold. Let us call M^n orientable whenever all its components are orientable, and nonorientable whenever it has a nonorientable component. Referring to (III, 19.1) we prove without difficulty that the restriction concerning connectedness may be omitted throughout.

II. Upon examining the theory developed in the present chapter, it will be found that the only properties of the linked complexes $Lk\ \sigma^p$ of K utilized anywhere are their homology groups. Therefore the results obtained, and notably Poincaré's duality theorem, will continue to hold if we replace the sphere property of the linked complexes by the following: $Lk\ \sigma^p$ has the homology groups of the $(n - p - 1)$-sphere, i.e. it is cyclic in the extreme dimensions and acyclic in the rest. A complex K with this property is known as a *combinatorial* manifold.

§3. Relative Homology Theory.

14. The relative homology theory and related developments offer much similarity with the "relative" concepts of point-sets. While this theory has many applications it shall be dealt with here very cursorily and largely to enable us to discuss relative manifolds and certain associated duality theorems.

Let $K = \{\sigma\}$ be a complex and L a subcomplex of K. Let us replace every chain C^p of K by its residue mod L, or equivalently identify two chains C^p, C'^p whenever their difference is in L. We shall also write accordingly

(14.1) $$C^p = C'^p \bmod L,$$

(14.2) $$FC^p = 0 \bmod L,$$

(14.3) $$C^p = FC^{p+1} \bmod L, \text{ or } C^p \sim 0 \bmod L.$$

When (14.2) holds C^p is called a *p-cycle of* K mod L or a *relative* cycle, and when (14.3) holds C^p is said to *bound* or to be *homologous* to zero mod L.

Let (K_1,L_1) be a pair such as (K,L) and Λ a linear chain-operator $K \to K_1$. If $C^p \subset K$ then we may write in unique manner

$$\Lambda C^p = \Lambda_1 C^p + \Lambda_2 C^p$$

where $\Lambda_1 C^p$ consists of the part of ΛC^p in $K_1 - L_1$ and $\Lambda_2 C^p$ of the part in L_1. Thus Λ_1 is a new linear chain-operator referred to as "Λ *reduced* mod L_1" or simply "Λ mod L_1."

If $\Lambda C^p \subset L_1$ for every $C^p \subset L$ we write $\Lambda L \subset L_1$.

Let us apply the preceding remark to F as an operator $K \to K$. In the first place $FL \subset L$. As a consequence if $F_1 = F$ mod L then

$$F_1F_1C^p = FFC^p \bmod L.$$

Hence $F_1F_1 = FF$ mod L and from $FF = 0$ follows

(14.4) $$F_1F_1 = 0.$$

This shows that a bounding chain mod L is a cycle mod L. As a consequence we may introduce for the relations mod L the same groups $\mathfrak{C}, \cdots, \mathfrak{H}$ as in the ordinary theory. The groups and the related Betti numbers are written $\mathfrak{C}^p(K,L;G), \cdots$. The cyclic terminology is also utilized in the customary way. The cycles, $\cdots$ of K itself are referred to as *absolute* cycles, $\cdots$.

It is clear that what has actually been done is to treat the set $K - L$ of simplexes like the set K, save that $F_1 = F$ mod L replaces F. It is therefore appropriate to refer to $K - L$ as a "complex" and more precisely as an *open* subcomplex of K. Accordingly L is known as a *closed* subcomplex of K. The justification of these terms rests upon property (15.2) below.

We shall revert to the use of the designation F (instead of F_1) for the boundary operator in $K - L$. Thus F is to be thought of as a universal boundary operator for all complexes.

The concepts: chain-mapping, homotopic chain-mappings, are

extended at once as between open subcomplexes. The meaning of the relation $C^p \subset K - L$ is also obvious.

For an intuitive illustration of the relative concepts we may have recourse as so often done before, to a planar disk Δ with some holes. Let Δ be covered with a complex K so that its borders are covered with a subcomplex L and that the arc λ is a subpath. Then λ is a one-cycle mod L or relative one-cycle. Notice also that every vertex is now ~ 0 mod L.

15. The following properties are immediate and their proofs are left to the reader:

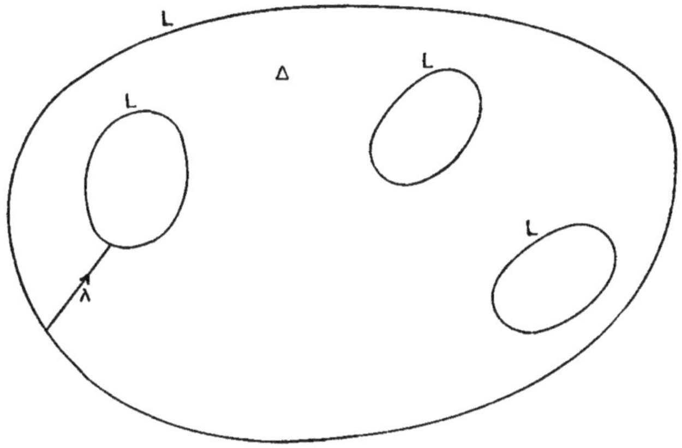

Figure 74

*(15.1) *A subset M of K is an open subcomplex if and only if it is true that if $\sigma \in M$ then St $\sigma \subset M$. In particular St σ is an open subcomplex.*

*(15.2) *If M is an open subcomplex of K then $|M|$ is an open subset of the polyhedron $|K|$.*

*(15.3) *$R^0_\pi(K,L)$ is independent of the integer π. It is written $R^0(K,L)$ and its value is the number of components of $|K|$ which do not meet $|L|$.*

*(15.4) *The Euler-Poincaré relation holds for $K - L$ and it reads*

$$\chi(K - L) = \Sigma(-1)^p\alpha^p = \Sigma(-1)^pR^p_\pi(K,L)$$

where α^p is the number of p-simplexes of $K - L$.

*(15.5) *An n-circuit* mod *L or "relative" n-circuit may be introduced and its treatment is the same as that of the ordinary or "absolute" n-circuit* (III, §7) *save that $K - L$ and relative cycles take the place of K and absolute cycles.*

16. Chain-mapping, subdivision, topological invariance. If τ is a chain-mapping $K \rightarrow K_1$ such that $\tau L \subset L_1$, then $\tau_1 = \tau$ reduced mod L_1 is readily verified to be a chain-mapping $K - L \rightarrow K_1 - L_1$.

Let τ, θ be chain-mappings $K \rightarrow K_1$ such that $\tau L \subset L_1$, $\theta L \subset L_1$ and let them be homotopic with operator $\mathfrak{D}$ such that $\mathfrak{D}L \subset L_1$. Passing to the reduced operators τ_1,θ_1 from the assumed relation due to $\tau \backsim \theta$:

$$\mathfrak{D}F + F\mathfrak{D} = \tau - \theta$$

follows

$$\mathfrak{D}_1F_1 + F_1\mathfrak{D}_1 = \tau_1 - \theta_1.$$

Hence $\tau_1 \backsim \theta_1$.

Let K_1 be a subdivision of K,L_1 the corresponding subdivision of L, d chain-subdivision $K \rightarrow K_1$ and d^{-1} an inverse of d. We verify at once that $dL \subset L_1$, $d^{-1}L_1 \subset L$. Furthermore if $\mathfrak{D}$ is the operator corresponding to the homotopy $dd^{-1} \backsim 1$ then $\mathfrak{D}L \subset L_1$. Hence if d_1,d_1^{-1} are d,d^{-1} reduced mod L_1 and mod L, from $d^{-1}d = 1$ and $dd^{-1} \backsim 1$ follows $d_1^{-1}d_1 = 1$ and $d_1d_1^{-1} \backsim 1$. Notice that $d_1 = d|(K - L)$. We refer to d_1 as subdivision in $K - L$ and to d_1^{-1} as an inverse of d_1. The relations just proved enable us to conclude as in (V, 6) that:

(16.1) *d_1 and d_1^{-1} induce isomorphisms of the homology groups of K* mod *L with the corresponding groups of K_1* mod *L_1.*

Let Π be any polyhedron, K a covering complex of Π, L a closed subcomplex of K. The polyhedron $\Omega = |L|$ is called a *subpolyhedron* of Π.

Given the polyhedron Π and the subpolyhedron Ω of Π, we may consider the class $\{K\}$ of the covering complexes such that for each K there is a closed subcomplex L of K covering Ω. The class of the pairs (K,L) has obvious topological character relative to the pair (Π,Ω).

An example of a polyhedron Π and a subpolyhedron Ω is an Euclidean sphere S^n, $n > 0$, and its equatorial sphere S^{n-1}.

By paraphrasing the treatment of (V, 7,8) we prove:

(16.2) *Let Π,Ω be a polyhedron and subpolyhedron and let K be a complex covering Π with a subcomplex L of K covering Ω. Then the homology groups of K* mod *L depend solely upon the pair* (Π,Ω) *and hence they are topological invariants of the pair (K,L). For this reason they are also referred to as homology groups of Π* mod *Ω, written* $\mathfrak{H}^p(\Pi,\Omega;G)$. *Similarly for the Betti numbers.*

For evident reasons we shall designate $\Pi - \Omega$ as an *open* polyhedron and by contrast Π and Ω as *closed* polyhedra.

17. The absolute cycles of an open polyhedron. We shall bring out the existence in an open polyhedron $\Pi - \Omega$ of a class of closed polyhedra which in a certain sense "approximate" $\Pi - \Omega$ arbitrarily closely, and in addition possess fixed homology groups. They will thus serve to define absolute cycles and groups for $\Pi - \Omega$. It is necessary however to discuss more fully the imbedding of Ω in Π.

Let $K = \{\sigma\}$ be a covering complex of Π such that a subcomplex L covers Ω. We shall refer to (K,L) as a *covering pair* for (Π,Ω).

The complex L is said to be *normal* in K, and the covering pair (K,L) is said to be *normal,* whenever any simplex of K whose vertices are all in L is necessarily itself in L. A vertex of K is manifestly normal in K.

Together with L we consider its star $St\ L$ or set of all the simplexes of K with a vertex in L and the complement $M = K - St\ L$ which is a closed subcomplex.

(17.1) *If K_1 is a subdivision of K and L_1 the induced subdivision of L then (K_1,L_1) is a normal covering pair. Thus normal covering pairs always exist and their collection has manifest topological character for* (Π,Ω).

If K',L' are the first derived of K,L it is manifestly sufficient to show that L' is normal in K'. Now if $\zeta = \hat{\sigma}_i \cdots \hat{\sigma}_j$, $\sigma_i < \cdots < \sigma_j$, is a simplex of K' with vertices all in L', then $\sigma_i, \cdots, \sigma_j$ must be simplexes of L and hence $\zeta \in L'$. Therefore L' is normal in K'.

(17.2) *If L is normal in K so is $M = K - St\ L$.*

If this property does not hold some $\sigma \in K$ has all its vertices in M without being in M. Therefore σ is in some star $St\ a$ of a vertex a of L. Hence a is a vertex of σ and consequently $a \in M$, which is ruled out since L and M are disjoint. Therefore (17.2) is true.

(17.3) *Let L be normal in K and set $R = K - L - M$. Through each point x of $|R|$ there passes a unique segment yz with y in $|L|$ and z in $|M|$. Moreover the transformation $\varphi : |St\ M| \to |M|$ defined by $\varphi x = z$ for $x \in |R|$, $\varphi x = x$ for $x \in |M|$ is a contraction.*

It may be observed that there is an analogous contraction $\psi : |St\ L| \to |L|$ but we shall not require it in the sequel.

The simplex of K containing x is neither in L nor in M, and hence it has vertices in both. Therefore it is of the form $\sigma\sigma'$, $\sigma \in L$, $\sigma' \in M$. Let $\sigma = a_o \cdots a_q$, $\sigma' = a_{q+1} \cdots a_p$, and let $t_o, \cdots, t_p$ be barycentric coordinates for $\sigma\sigma'$. Then, for the point x, the numbers

$$\lambda = \sum_{i \leqq q} t_i, \qquad \mu = \sum_{i > q} t_i,$$

$$s_i = t_i/\lambda, \quad i \leq q; \qquad s_{q+1} = 0,$$

$$s_i' = 0, \quad i \leq q; \qquad s_{q+i}' = t_{q+1}/\mu$$

are uniquely determined with $\lambda > 0$, $\mu > 0$. In our usual designations the s_i are the barycentric coordinates of a point y of σ, the s_i' those of a point z of σ' such that $x = \lambda y + \mu z$, $\lambda + \mu = 1$. Therefore x is in the segment yz.

Suppose that there is a second segment y_1z_1, $y_1 \in \sigma$, $z_1 \in \sigma'$, containing x, where $y_1 \neq y$. As a consequence $x = \lambda y + \mu z = \lambda_1 y_1 + \mu_1 z_1$. Hence expressing y,z,y_1,z_1 in terms of $a_o, \cdots, a_p$, there is obtained a relation

$$\nu_o a_o + \cdots + \nu_p a_p = 0$$

where, owing to $y_1 \neq y$, one of the coefficients $\nu_o, \cdots, \nu_p$ is not zero. Since the vertices are independent this is excluded, and so the segment yz is unique.

Since the barycentric coordinates of z are continuous functions of those of x, φ is a mapping. The existence of the segments xz whenever $x \neq z$, i.e. $x \in |R|$, shows that φ is a contraction $|St\ M| \to |M|$.

18. (18.1) **Theorem.** *Let* (K,L) *be a normal covering pair of* (Π,Ω) *and* $M = K - St\ L$. *Then the homology groups of* M *are topological invariants of the open polyhedron* $\Pi - \Omega$.

That is to say if K^*,L^* is another normal couple and $M^* = K^* - St\ L^*$ then M,M^* have isomorphic homology groups. These groups give rise then to what we shall designate as the *absolute homology groups* of $\Pi - \Omega$. The term is entirely appropriate since the cycles dealt with are absolute cycles of M,M^*. We must prove:

(18.2) *The corresponding homology groups of* M *and* M^* *are isomorphic.*

If K_1 is a subdivision of K let L_1 denote the associated subdivision of L and set $M_1 = K_1 - St\ L_1$, $B_1 = M_1 \cap Cl\ St\ L_1$. Similar notations are used for K^*. It is clear that $|St\ L_1| \subset |St\ L|$, hence $|M_1| \supset |M|$. Moreover the distance from $|L|$ to $|M_1|$ is positive and less than the mesh of K_1. Hence it may be chosen arbitrarily small.

Let λ denote the segment yz of (17.3). Each λ meets $|B_1|$ in a single point. This is elementary when K_1 is the first derived K', and holds by repetition for an arbitrary K_1. It follows that φ induces a deformation (contraction) $|M_1| \to |M|$. The induced operation on the homology groups will be written M_1M and as we know (V, 9.4) it is an isomorphism. Let us first identify all the homology classes of M or its subdivisions corresponding to one another under chain-subdivision, and similarly for M_1. Let us then identify the classes which correspond under isomorphisms of the type M_1M. This causes an identification of a large collection of homology groups, associated with K. The resulting groups and elements are written temporarily $\mathfrak{H}^p(G)$, Γ^p, and called as usual homology groups and classes. A cycle γ^p in one of the classes identified to form Γ^p is referred to as a *representative* of Γ^p. Under these new conventions $M_1M = 1$.

A similar treatment is applied to K^* and M^* resulting in $\mathfrak{H}^{*p}(G)$, Γ^{*p}.

As regards our theorem we may replace K^* by any subdivision. Hence we may suppose $|M| \subset |M^*| \subset |M_1| \subset |M_1^*|$. The identity mappings $|M| \to |M^*|, \cdots$ give rise to homomorphisms of the homology groups $\mathfrak{H}^p(G), \cdots$, which we denote by $MM^*, \cdots$ and clearly $M^*M_1 \cdot MM^* = MM_1$. Let M_{11} denote the subdivision of M induced by K_1, so that M_{11} is a subcomplex of M_1. Since $M_1M = 1$, every Γ^p has a representative in M and hence a representative γ^p in M_{11}. It is clear that γ^p is a representative in M_1 and as it is mapped into itself as a consequence of the imbedding of M into M_1, $MM_1 = 1$. Therefore

$$M^*M_1 \cdot MM^* = 1.$$

Similarly

$$M_1M_1^* \cdot M^*M_1 = 1.$$

Hence M^*M_1 is an isomorphism between $\mathfrak{H}^p(G)$ and $\mathfrak{H}^{*p}(G)$. Using this and similar isomorphisms we may identify all the homology groups $\mathfrak{H}^p(M,G)$ into a single homology group $\mathfrak{H}^p(\Pi - \Omega, G)$ and this more than proves theorem (18.1).

If Γ^p is any element (homology class) of one of the final groups and γ^p a cycle of M or M_{11} in one of the identified classes, we shall refer to γ^p as a *representative* of Γ^p. We shall also call, less accurately, γ^p and Γ^p an *absolute cycle* and *homology class* of $\Pi - \Omega$.

19. Relation to connectedness. It is exactly the same for the open polyhedron $\Pi - \Omega$ as for the closed polyhedron Π. More explicitly:

(19.1) *The number $R_\pi^o(\Pi - \Omega) = R_\pi^o(M)$ is independent of the integer π and its fixed value, written $R^o(\Pi - \Omega)$, is equal to the number of components of $\Pi - \Omega$.*

Suppose that $\{U_1, \cdots, U_r\}$ is a finite set of distinct components of $\Pi - \Omega$. Since $|M_1|$ of (18) approximates $\Pi - \Omega$ arbitrarily closely it may be chosen such as to meet each of the components. The intersections $|M_1| \cap U_i$ are distinct components of $|M_i|$. Hence $r \leq R^o(M_1) = R^o(M)$. Therefore the number ρ of components is finite and

$$\rho \leq R^o(M). \tag{19.2}$$

Suppose that in (19.2) the equality does not hold. Then $|M|$ has more components than ρ and so there exist in $|M|$ two points x,y which are in no connected subset of $|M|$ but in some connected subset λ (in fact an arc) of $\Pi - \Omega$. Applying the contraction φ of (17.3) there is obtained a connected subset $\varphi\lambda$ of $|M|$ containing both points. This contradiction shows that the equality alone holds in (19.2) and this proves (19.1).

Example: $R^o(\sigma^n) = 1$ since σ^n is connected.

§4. RELATIVE MANIFOLDS AND RELATED DUALITY THEORY (ELEMENTARY THEORY). ALEXANDER'S DUALITY THEOREM.

20. With the means at our disposal we cannot go very far and in particular we can only treat very special cases of the duality theorems. Thus we shall have to assume throughout that we are dealing with open or closed polyhedra, a limitation due exclusively to the weakness of our methods. On the other hand it must be said that the restrictions imposed will enable us to obtain very simple and rapid proofs of the main results: the author's relative duality theorems and Alexander's duality theorem.

Let M^n be an orientable differentiable manifold, and Π a subpolyhedron of M^n. We refer to the open polyhedron $M^n - \Pi$ as an *open* manifold, or a *relative* manifold, or also a *manifold* mod Π. By contrast M^n itself is referred to as *absolute.*

There exist then by hypothesis a covering normal pair (K,L): the complex K covers M^n and the subcomplex L covers Π and is normal in K.

Examples. A segment is a one-manifold modulo the set of its two end-points. A closed surface Φ in ordinary space bounds a region Ω and Ω is a three-manifold mod Φ. If Δ^{n+1} is an $(n+1)$-disk with boundary sphere S^n, then $\Delta^{n+1} - S^n$ is an $(n+1)$-manifold mod S^n.

Reciprocation calls for some special remarks. It will be necessary first of all to deal with subcomplexes of the reciprocal $\bar{K}$ of K. Such a subcomplex $\bar{K}_1$ is merely a subset in which the same incidence relations as in $\bar{K}$ still yield $FF = 0$. In particular $\bar{K}_1$ is a subcomplex whenever one of the following two conditions holds:

(a) If $\xi \in \bar{K}_1$ then $Cl\,\xi \subset \bar{K}_1$. Such a subcomplex is said to be *closed.*

(b) If $\xi \in \bar{K}_1$ then $St\,\xi \subset \bar{K}_1$. Such a subcomplex is said to be *open.*

If K_1 is a subcomplex of K it is readily seen that the set $\bar{K}_1$ of the reciprocals of the elements of K_1 is a subcomplex of $\bar{K}$. Derivation in $\bar{K}$ transforms $\bar{K}_1$ into a set written $\bar{K}_1'$ and called the *derived* of $\bar{K}_1$.

(20.1) *Let K,L be as before. Then $\bar{L}$ is an open subcomplex of $\bar{K}$ and* *the derived* $(\bar{L})'$ *coincides with* $St\, L'$ *in* K'. *Hence* $\overline{K - L} = \bar{K} - \bar{L}$ *is a closed subcomplex of* $\bar{K}$ *and* $(\overline{K - L})' = K' - St\, L'$.

Suppose $\xi_i^{n-p} \in \bar{L}$ so that $\sigma_i^p \in L$. Then $\bar{\delta}\xi_i^{n-p}$ consists of all the simplexes $\zeta = \hat{\sigma}_i^p \hat{\sigma}_j^{p+1} \cdots \hat{\sigma}_h^n$, $\sigma_i^p < \cdots < \sigma_h^n$. These simplexes and all their faces beginning with $\hat{\sigma}_i^p$ make up all the simplexes beginning with a vertex $\hat{\sigma}$ such that $\sigma \in L$. Their totality is therefore $St\, L'$ in K' and so $(\bar{L})' = St\, L'$. If $\xi_i^{n-p} < \xi_j^{n-q}$ then $\sigma_j^q < \sigma_i^p$. Hence σ_j^q is in L and therefore ξ_j^{n-q} is in $\bar{L}$. Thus if $\xi \in \bar{L}$ then $St\, \xi \subset \bar{L}$ and so $\bar{L}$ is an open subcomplex of $\bar{K}$. This proves (20.1).

Since $\overline{K - L}$ is a closed subcomplex of $\bar{K}$ its relation to its derived may be developed as for $\bar{K}$. Hence the homology groups of $\overline{K - L}$ and of $K' - St\, L' = (\overline{K - L})'$ are isomorphic. Coupling this with (18.1), and recalling that $K' - St\, L'$ is the complex M of (17) we have:

(20.2) *The homology groups of* $\overline{K - L}$ *are isomorphic with those of* $M^n - \Pi$.

Regarding the relations between $K - L$ and $\overline{K - L}$ the same arguments may be developed as before for the absolute manifold. However, circuits and cycles must now be taken mod L in K, but the cycles continue to be absolute in $\overline{K - L}$. As a consequence we find instead of (7.2) the relation

$$R_\pi^p(M^n, \Pi) = R_\pi^{n-p}(\overline{K - L}).$$

From this follows the author's generalization of Poincaré's duality theorem:

(20.3) *Duality theorem for relative manifolds. Let* M^n *be an orientable manifold and* Π *a subpolyhedron of* M^n. *Then the Betti numbers of the cycles of* $M^n - \Pi$ *and of* M^n mod Π *satisfy the duality relation:*

$$R_\pi^p(M^n - \Pi) = R_\pi^{n-p}(M^n, \Pi), \tag{20.4}$$

$$(\pi = 0 \textit{ for rational cycles}).$$

When M^n *is nonorientable one may only assert that*

$$R_2^p(M^n - \Pi) = R_2^{n-p}(M^n, \Pi). \tag{20.5}$$

21. Manifolds with regular boundary. Intuitively this is a very natural concept. However if one attempts to be quite general the definition becomes anything but simple. To avoid all complications let us start with the same differentiable orientable M^n as before and consider in M^n a region (open set) Ω whose boundary consists of a finite set $M_1^{n-1}, \cdots, M_r^{n-1}$ of disjoint closed orientable differentiable manifolds, each regularly imbedded in M^n, and with the following additional property: Each M_i^{n-1} has a neighborhood N_i in $\bar{\Omega}$ such that $\bar{N}_i$ is homeomorphic with a product $l \times M_i^{n-1}$ where l is a segment. Under the circumstances Ω is known as an *orientable n-manifold with regular boundary.* The boundary $\mathfrak{B}(\Omega) = \bigcup M_i^{n-1}$ will be denoted by B. We shall assume throughout that B is a subpolyhedron of M^n. Thus there are covering complexes K with subcomplexes covering B. It may be said that all the assumptions are readily verified in the simple cases.

Since the M_i^{n-1} are compact and disjoint their distances apart have a positive lower bound. Hence one may choose the N_i disjoint. This is assumed in what follows.

Each point $x \in M_i^{n-1}$ is the end-point of a unique arc λ which is the image of l in $\bar{N}_i$. Let y be the other end-point of λ. Let a mapping $\varphi : \bar{\Omega} \to \Omega$ be defined by the condition that each arc λ is shortened in a certain ratio $\rho < 1$ from yx to yz, while φ is the identity outside of the N_i. It is clear that φ is a topological mapping of $\bar{\Omega}$ into a subset $\bar{\Omega}_\rho$ of $\bar{\Omega}$. By reasonably evident paraphrase of (18) with the sets $\bar{\Omega}_\rho$ taking the place of the complexes M, we may prove:

(21.1) *The homology groups of* $\bar{\Omega}$ *are isomorphic with the corresponding groups of* Ω.

By combining with (20.3) and writing M^n for $\bar{\Omega}$ there is obtained the author's

(21.2) *Duality theorem for manifolds with regular boundary. Let* M^n *be an orientable n-manifold with regular boundary B. Then there takes place the duality relation*

$$R_\pi^p(M^n) = R_\pi^{n-p}(M^n,B) \tag{21.3}$$

$$(\pi = 0 : \textit{rational cycles}).$$

When the manifolds are nonorientable one may only assert that:

$$R_2^p(M^n) = R_2^{n-p}(M^n,B). \tag{21.4}$$

Examples. Let Δ be a plane disk with p holes whose borders are circumferences. Then Δ is an orientable M^2 with regular boundary B consisting of $p + 1$ circumferences. The number $R_\pi^1(M^2) = p$ whatever π. Hence $R_\pi^1(M,B) = p$. In fact if we draw p arcs joining the exterior circumference to each interior circumference, these p arcs give rise to p independent one-cycles mod B which form a base mod π for all one-cycles mod B. Since $R^0(M^2) = 1$ we have $R^2(M^2,B) = 1$, a fact readily verified directly.

It may be noticed that all the examples of (20) are cases of manifolds with regular boundaries.

22. The Alexander duality theorem. Let S^n be an n-sphere and Π a subpolyhedron of S^n. That is to say it is assumed that the pair (S^n,Π) has a normal covering pair (K,L). Then according to Alexander

(22.1) **Theorem.** *Between the Betti numbers of the closed polyhedron* Π *and those of the open polyhedron* $S^n - \Pi$ *there takes place the duality relation*

$$R^p_\pi(S^n - \Pi) = R^{n-p-1}_\pi(\Pi) + \delta_{po} - \delta_{p,n-1} \tag{22.2}$$

$$(\pi = 0 : \textit{rational cycles}),$$

where the δ_{ij} *are the Kronecker deltas.*

Coupling (22.1) with (19.1) we obtain the

(22.3) **Corollary.** *The number of components of* $S^n - \Pi$ *is* $1 + R^{n-1}_\pi(\Pi)$. *Hence this last Betti number is independent of* π *and is written* $R^{n-1}(\Pi)$.

Example. If J is a Jordan curve in S^2 then the number of components is $1 + R^1(J) = 2$, as was to be expected.

Proof of (22.1). From (20.4) there follows

$$R^p_\pi(S^n - \Pi) = R^{n-p}_\pi(S^n,\Pi). \tag{22.4}$$

Replacing $n - p$ by $p + 1$ this reduces the proof to

$$(22.5)_p \qquad R^{p+1}_\pi(S^n,\Pi) = R^p_\pi(\Pi) + \delta_{n-p-1,0} - \delta_{n-p-1,n-1}.$$

We dismiss at the outset the special cases $n = 1$ and $\Pi = S^n$: they are easily dealt with directly and in addition they are of little interest. We assume then $\Pi \neq S^n$ and $n > 1$. Since L will now be a proper subcomplex of the n-circuit K,L will contain no n-cycle and clearly we may also assume $p < n$.

The following designations will be used: γ and C for a cycle and a chain of K, δ and D for a cycle and a chain of L.

Suppose first $0 < p < n - 1$. If γ^{p+1} is a cycle of K mod L then $\delta^p = F\gamma^{p+1}$ is a cycle of L. If $\gamma^{p+1} \sim 0$ mod L then $\gamma^{p+1} = FC^{p+2} + D^{p+1}$, hence $\delta^p = F\gamma^{p+1} = FD^{p+1} \sim 0$ in L. Hence $\gamma^{p+1} \to \delta^p$ defines a homomorphism θ of the homology group of the $(p + 1)$-cycles mod L and mod π, into the (absolute) homology group of the p-cycles of L mod π. If $\delta^p \sim 0$ in L then $\delta^p = FD^{p+1}$ and $\gamma^{p+1} - D^{p+1}$ is a cycle of K. Since $0 < p + 1 < n$, this cycle ~ 0 in K, and therefore $\gamma^{p+1} \sim 0$ mod L. Thus θ is univalent. If δ^p is any cycle of L and since $0 < p < n - 1$, $\delta^p \sim 0$ in K, or $\delta^p = F\gamma^{p+1}$. Thus γ^{p+1} is a cycle of K mod L whose existence proves that θ is onto. Thus θ is an isomorphism of the homology group of the $(p + 1)$-cycles mod L and mod π with the homology group of the p-cycles mod π of L. The resulting equality of the Betti numbers is $(22.5)_p$ which is thus proved for the range under consideration.

Suppose now $p = n - 1$. The result is exactly the same as before provided that we do not distinguish between two cycles γ^n, γ'^n mod L which differ by a multiple of the fundamental n-cycle of K. Thus $R^n - 1$ takes the place of R^{p+1} and $(22.5)_{n-1}$ follows.

Suppose finally $p = 0$. This time the reasoning is again the same provided that the only zero-cycles considered in L are those ~ 0 in K, and these are the cycles with Kronecker index zero. This necessitates replacing $R^o_\pi(\Pi)$ by $R^o_\pi(\Pi) - 1$ and yields $(22.5)_o$.

This completes the proof of Alexander's duality theorem.

Remark. We repeat our earlier observation: all the duality theorems here proved are valid under much more general conditions. It may be stated however that they are not changed in form as long as the discourse is confined to Betti numbers.

PROBLEMS

1. Let $f(x_1, \cdots, x_{n+1})$, $n > 0$, be a real polynomial and let Φ be the locus of $\mathfrak{E}^{n+1}$ represented by $f = 0$. A point of Φ where all the partial derivatives $\frac{\partial f}{\partial x_i}$ vanish is called a *singular point* of Φ. Show that a bounded component of Φ without singular points is a compact orientable M^n. Prove that this M^n is a polyhedron (special case of the theorem of Cairns).

2. The situation being as in (V, problem 4) show that a complex projective plane is a compact orientable M^4.

3. The notations being again those of (V, problem 4) if $f(x,y,z)$ is a homogeneous complex polynomial the complex locus $\Gamma : f = 0$ is known as an *algebraic curve.* A point of Γ where the three partials $\frac{\partial f}{\partial x}, \cdots,$ vanish is called a *singular point* of Γ. Show that if Γ has no singular points and f is irreducible then Γ is a compact orientable M^2. Generalize.

4. Show that the algebraic curve Γ represented by $y^2z = x^3 + pxz^2 + qz^3$, (where $x^3 + px + q = 0$ has no multiple roots), is a surface of genus one.

5. Let M^p, M^q be compact orientable manifolds. Then $M^p \times M^q$ is a compact orientable M^{p+q}.

6. Let $K,\bar{K},\sigma,\xi$ have the same meaning as in (9) for M^n and let $K_1, \cdots$ be the analogues for a second manifold M^n_1. If

$$\tau\sigma^p_i = \Sigma a_{ij}(p)\sigma^p_{1j}$$

is a chain-mapping $K \rightarrow K_1$ then

$$\bar{\tau}\xi_{ij}^{n-p} = \Sigma a_{ij}(p)\xi_i^{n-p}$$

is likewise a chain-mapping $\bar{K}_1 \rightarrow \bar{K}$. It is called the *reciprocal* of τ. (Hopf).

7. Let M^n be a compact orientable manifold and S^n an n-sphere. Prove that a n.a.s.c. for two mappings $M^n \rightarrow S^n$ to be homotopic is that they have the same degree.

8. Let Π be a polyhedron and x any point of Π. In any complex $K = \{\sigma\}$ covering Π let $\sigma(x)$ denote the simplex containing x. Prove that the homology groups of the star $St\ \sigma(x)$ are topologically invariant (independent of K). They are called the *local groups* of x.

9. Prove by means of the local groups that an n-circuit is topologically invariant. That is to say if Π is covered by a complex K which is an n-circuit then every such complex has the same property. Similarly for the orientable n-circuit.

10. Prove by means of the local groups the invariance of dimensionality theorems of Brouwer of (IV, 12).

Bibliography

The present bibliography, admittedly very sketchy, is merely meant to give the beginner in topology a short list of easily accessible important books and papers on the topics treated in the text. Much more extensive bibliographies will be found in the author's two volumes of the Colloquium Series.

LIST OF BOOKS

ALEXANDROFF, P., and HOPF, H.

[A-H] *Topologie*, I, Berlin, Springer, 1935, (Grundlehren der mathematischen Wissenschaften, bd. 45).

HUREWICZ, W., and WALLMAN, H.

[H-W] *Dimension Theory*, Princeton University Press, 1941, (Princeton Mathematical Series, no. 4).

KERÉKJÁRTÓ, B.

[K] *Vorlesungen über Topologie*, Berlin, Springer, 1923, (Grundlehren der mathematischen Wissenschaften, bd. 8).

LEFSCHETZ, S.

[L_1] *Topology*, New York, 1930, (American Mathematical Society, Colloquium Publications, vol. 12).

[L_2] *Algebraic Topology*, New York, 1942, (American Mathematical Society, Colloquium Publications, vol. 27).

SEIFERT, H., and THRELFALL, W.

[S-T] *Lehrbuch der Topologie*, Berlin, Teubner, 1934.

VEBLEN, O.

[V] *Analysis Situs*, 2d edition, New York, 1931, (American Mathematical Society, Colloquium Publications, vol. 5, part 2).

WILDER, R. L., editor.

[UM] *Lectures in Topology*, University of Michigan Press, 1941.

LIST OF PAPERS

ALEXANDER, J. W.

[a] *A proof of the invariance of certain constants of analysis situs*, Transactions, American Mathematical Society, vol. 16 (1915), pp. 148–154.

[b] *A proof and extension of the Jordan-Brouwer separation theorem*, Transactions, American Mathematical Society, vol. 23 (1922), pp. 333–349.

[c] *Combinatorial analysis situs*, I, Transactions, American Mathematical Society, vol. 28 (1926), pp. 301–329.

BROUWER, L. E. J.

[a] *On continuous vector distributions on surfaces*, Proceedings, Akademie van Wetenschappen, Amsterdam, vol. 11, part 2 (1909), pp. 850–858; vol. 12 (1910), pp. 716–734; vol. 13, part 1 (1910), pp. 171–186.

[b] *On continuous one-to-one transformations of surfaces into themselves*, Proceedings, Akademie van Wetenschappen, Amsterdam, vol. 11, part 2 (1909), pp. 788–798; vol. 12 (1910), pp. 286–297; vol. 13, part 2 (1911), pp. 767–777; vol. 14, part 1 (1911), pp. 300–310; vol. 15, part 1 (1913), pp. 352–360; vol. 22, part 2 (1920), pp. 811–814; vol. 23, part 1 (1921), pp. 232–234.

[c] *Zur analysis situs*, Mathematische Annalen, vol. 68 (1910), pp. 422–434.

[d] *Beweis der Invarianz der Dimensionenzahl*, Mathematische Annalen, vol. 70 (1911), pp. 161–165.

[e] *Über Abbildungen von Mannigfaltigkeiten*, Mathematische Annalen, vol. 71 (1912), pp. 97–115.

[f] *Invarianz des n-dimensionalen Gebiets*, Mathematische Annalen, vol. 71, pp. 305–313; vol. 72, pp. 55–56 (1912).

CAIRNS, S. S.

[a] *Transformation of regular loci*, Annals of Mathematics, (2), vol. 35 (1934), pp. 579–587.

EILENBERG, S.

[a] *Extension and classification of continuous mappings*, see under Wilder [UM] (1941), pp. 57–99.

HOPF, H.

[a] *Abbildungsklassen n-dimensionaler Mannigfaltigkeiten*, Mathematische Annalen, vol. 96 (1927), pp. 209–224.

[b] *Vektorfelder in n-dimensionalen Mannigfaltigkeiten*, Mathematische Annalen, vol. 96 (1927), pp. 225–250.

[c] *Mindestzahlen von Fixpunkten*, Mathematische Zeitschrift, vol. 26 (1927), pp. 762–774.

[d] *New proof of the Lefschetz formula on invariant points*, Proceedings, National Academy of Sciences, vol. 14 (1928), pp. 149–153.

[e] *Verallgemeinerung der Euler-Poincaréschen Formel*, Gesellschaft der wissenschaften zu Göttingen, Mathematischphysikalische Klasse., Nachrichten, 1928, pp. 127–136.

[f] *Topologie der Abbildungen von Mannigfaltigkeiten. I. Neue Darstellung der Theorie des Abbildungsgrades*, Mathematische Annalen, vol. 100 (1928), pp. 579–608; *II. Klasseninvarianten*, Mathematische Annalen, vol. 102 (1930), pp. 562–623.

HUREWICZ, W.

[a] *Beiträge zur Topologie der Deformationen. I. Höherdimensionale Homotopiegruppen*, Proceedings, Akademie van Wetenschappen, Amsterdam, vol. 38, part 1 (1935), pp. 112–119; *II. Homotopie und Homologiegruppen*, ibid., pp. 521–528; *III. Klassen und Homologietypen von Abbildungen*, vol. 39, part 1 (1936), pp. 117–126; *IV. Asphärische Räume*, ibid., pp. 215–224.

LEFSCHETZ, S.

[a] *Intersections and transformations of complexes and manifolds*, Transactions, American Mathematical Society, vol. 28 (1926), pp. 1–49.

[b] *Manifolds with a boundary and their transformations*, Transactions, American Mathematical Society, vol. 29 (1927), pp. 429–462.

POINCARÉ, H.

[a] *Analysis situs*, Journal de l'École Polytechnique, Paris, (2), vol. 1

(1895), pp. 1–123. *Complément,* Rendiconti, Circolo Matematico, Palermo, vol. 13 (1899), pp. 285–343. *Deuxième complément,* Proceedings, London Mathematical Society, vol. 32 (1900), pp. 277–308. *Cinquième complément,* Rendiconti, Circolo Matematico, Palermo, vol. 18 (1904), pp. 45–110.

PONTRJAGIN, L.

[a] *Mapping of the three-dimensional sphere into an n-dimensional complex,* Akademiia nauk, SSSR, Doklady, vol. 34 (1942), pp. 35–37. *Characteristic cycles,* ibid., vol. 47 (1945), pp. 242–245. *Classification of some skew products,* ibid., pp. 322–325.

STEENROD, N. E.

[a] *Topological methods for the construction of tensor functions,* Annals of Mathematics, vol. 43 (1942), pp. 116–131. *Classification of sphere bundles,* ibid., vol. 45 (1944), pp. 294–311. *Homotopy relations in fibre spaces,* with W. Hurewicz, Proceedings, National Academy of Sciences, vol. 27 (1941), pp. 60–64.

TUCKER, A. W.

[a] *An abstract approach to manifolds,* Annals of Mathematics, (2), vol. 34 (1933), pp. 191–243.

[b] *Some topological properties of disk and sphere,* Proceedings, Canadian Mathematical Congress (Montreal 1945), pp. 285–309.

WHITEHEAD, G. W.

[a] *Homotopy properties of the real orthogonal groups,* Annals of Mathematics, vol. 43 (1942), pp. 132–146. *Homotopy groups of spheres and rotation groups,* ibid., pp. 634–640. *A generalization of the Hopf invariant,* Proceedings, National Academy of Sciences, vol. 32 (1946), pp. 188–190. *Products in homotopy groups,* Annals of Mathematics, vol. 47 (1946), pp. 460–475.

WHITEHEAD, J. H. C.

[a] *On C^1-complexes,* Annals of Mathematics, (2), vol. 41 (1940), pp. 809–824.

[b] *Simplicial spaces, nuclei, and m-groups,* Proceedings, London Mathematical Society, (2), vol. 45 (1939), pp. 243–347. *On adding relations to homotopy groups,* Annals of Mathematics, vol. 42 (1941), pp. 409–428. *Incidence matrices, nuclei, and homotopy type,* ibid., pp. 1197–1239. *Groups $\pi_r(V_{n,m})$ and sphere bundles,* Proceedings, London Mathematical Society, (2), vol. 48 (1944), pp. 243–291. *Note on a previous paper entitled "On adding relations to homotopy groups,"* Annals of Mathematics, vol. 47 (1946), pp. 806–810.

WHITNEY, H.

[a] *Differentiable manifolds,* Annals of Mathematics, (2), vol. 37 (1936), pp. 645–680. *Topology of differentiable manifolds,* see under Wilder [UM] (1941), pp. 101–143.

WILDER, R. L.

[a] *The sphere in topology,* American Mathematical Society, Semicentennial Publications, vol. 2 (1938), pp. 136–184.

List of Symbols

Each symbol is listed with the number of the page where it is introduced and explained.

Index

Milton Keynes UK
Ingram Content Group UK Ltd.
UKHW010024170724
445591UK00005B/168